Generation of Self-Excited, Hidden and Non-Self-Excited Attractors in Piecewise Linear Systems

Some Recent Approaches

Generation of Self-Excited, Hidden and Non-Self-Excited Attractors in Piecewise Linear Systems

Some Recent Approaches

Eric Campos Cantón
Instituto Potosino de Investigación Científica y Tecnológica A.C., Mexico

Rodolfo de Jesús Escalante González
Technological Institute of San Luís Potosí, Mexico

Héctor Eduardo Gilardi Velázquez
Universidad Panamericana, Aguascalientes, Mexico

World Scientific

NEW JERSEY • LONDON • SINGAPORE • BEIJING • SHANGHAI • HONG KONG • TAIPEI • CHENNAI • TOKYO

Published by

World Scientific Publishing Co. Pte. Ltd.
5 Toh Tuck Link, Singapore 596224
USA office: 27 Warren Street, Suite 401-402, Hackensack, NJ 07601
UK office: 57 Shelton Street, Covent Garden, London WC2H 9HE

Library of Congress Cataloging-in-Publication Data
Names: Campos Cantón, Eric author. | Escalante González, Rodolfo de Jesús, author. |
    Gilardi Velázquez, Héctor Eduardo, author.
Title: Generation of self-excited, hidden and non-self-excited attractors in
    piecewise linear systems : some recent approaches / Eric Campos Cantón,
    Rodolfo de Jesús Escalante González, Héctor Eduardo Gilardi Velázquez.
Description: New Jersey : World Scientific, [2023] | Includes bibliographical references and index.
Identifiers: LCCN 2023008262 | ISBN 9789811274114 (hardcover) |
    ISBN 9789811274121 (ebook) | ISBN 9789811274138 (ebook other)
Subjects: LCSH: Linear systems. | Piecewise linear topology. | Attractors (Mathematics) |
    Chaotic behavior in systems.
Classification: LCC QA402 .C356 2023 | DDC 003--dc23/eng20230506
LC record available at https://lccn.loc.gov/2023008262

British Library Cataloguing-in-Publication Data
A catalogue record for this book is available from the British Library.

For any available supplementary material, please visit
https://www.worldscientific.com/worldscibooks/10.1142/13347#t=suppl

Typeset by Stallion Press
Email: enquiries@stallionpress.com

Preface

Human beings have always had the desire to understand the behavior of the universe, we have developed different techniques and tools for the generation of knowledge structures that have allowed us to understand different phenomena in nature, seeking to formulate laws that often describe complex behavior. As regards the exact sciences, mathematical models have been developed that help us to describe the temporal evolution of the universe, either deterministically or stochastically. Under certain conditions a dynamical system can be deterministic and not be predictable. Despite the complexity that a system may have, on many occasions, it can be broken down into its simplest parts and analyzed in this way; however, in some of these cases there are limitations that do not allow us to predict their behavior or build a deterministic dynamic system that represents it. This does not mean that something cannot be said about such systems due to the properties they present. For these systems can be understood with the help of chaos theory, an example of which are turbulences and physical-chemical and biological oscillations.

There is a great diversity of dynamical systems dedicated to understanding the mechanism of generation of chaotic attractors, and more and more recent results are offered for advanced undergraduate students and beginning graduate students. This book covers those topic necessary for a clear understanding of the generation of self-excited, hidden and non-self-excited attractors. These three classes of attractors are addressed with piecewise linear systems. The fundamentals of linear theory for continuous-time dynamical systems, Lyapunov exponents and chaos concept are presented in Chapter 1.

Many interesting concepts in this book are given for piecewise linear systems for the generation of attractors. Some approaches to generate the attractors are presented such as saturation technique, step function, hysteresis function. Most of this book is devoted to the study of a class of systems that are dissipative and unstable, this type of system is known as a UDS system for its acronym for Unstable Dissipative System. Self-excited chaotic attractors occur through heteroclinic orbits, starting from the double-scroll attractor to the multiscroll sttractos. One goal is to describe two mechanisms of the generation of two-directional and three-directional multi-scroll attractors. This is done in Chapter 2, which deals with the concepts from the point of view of self-excited attractors. Also, in this Chapter 2, two important issues of behaviors in dynamical systems are developed: multi-stable systems and hyperchaotic systems.

Currently, hidden and non-self-excited attractors are still an interesting topic. Although a large number of systems with these kind of attractors have been reported in the last two decades, several questions related to their existence and characteristics remain open. One thing to consider is that the chaotic behavior exhibited by these attractors cannot be explained through the equilibria-based theory developed for self-excited attractors. Therefore, the study of these classes of attractors could eventually lead to ideas that could also be applied to self-excited attractors and various dynamical systems. The advantages of using these type of attractors in different applications are also still under study.

Chapter 3 focuses on these two classes of attractors, some reported systems with hidden and non-self-excited attractors are presented, as well as some recent approaches that allow the design of systems with coexistence of chaotic attractors. Chapter 3 addresses two types of multistability: coexistence of non-self-excited attractors and coexistence of self-excited attractors along with hidden attractors. An approach for the design of systems with non-self-excited hyperchaotic attractors is also presented as well as design techniques for the generation of one-directional, two-directional and three-directional scroll attractors of these classes.

Derivatives and integrals of fractional-order are generalizations of those of integer order; with the implementation of fractional calculus theory, developed almost 300 years ago, the study of chaotic systems through fractional-order-derivatives, has been extensively addressed by the scientific community. These studies have given possibility of finding new behaviors and better descriptions of natural phenomena that have been a recurring theme in the literature. In the same way as in integer-order systems, numerous publications describe the implementation of chaotic systems under fractional-order schemes. Chapter 4 is devoted to fractional

Piecewise Linear (PWL) Systems, first some bases of fractional calculus are presented, as well as the effects caused by that the fractional operators in scroll attractors. A comparison of the effects produced in the dynamics of chaotic system under the use of fractional-order with its integer counterpart is discussed. The intrinsic properties analyzed are correlations, Poincare sections, and system response frequency.

E. Campos-Cantón
R. de J. Escalante-González
H. E. Gilardi-Velázquez

Contents

Chapter 1

Introduction to dynamical systems and chaos

1.1 Continuous time dynamical systems

Dynamical systems are a branch of mathematics dedicated to understanding processes in motion, such processes occur in all branches of science. For instance, there is a great diversity of dynamical systems such as the motion of stars and galaxies, the stock market, the weather, chemical reactions, population growth, the motion of a simple pendulum, among other. We can clearly see that some dynamical systems are predictable while others are not. The evolution of the state of a dynamical system in time can be in continuous time ($t \in \mathbb{R}$) or discrete time ($t \in \mathbb{Z}$). For example, differential equations evolve in continuous time

$$\dot{\mathbf{x}} = f(\mathbf{x}), \text{ with } t \in \mathbb{R}, \tag{1.1}$$

where $\dot{\mathbf{x}} = d\mathbf{x}/dt$ and $\mathbf{x} \in \mathbb{R}^n$ is the state vector of the system. While, difference equations evolve in discrete time

$$\mathbf{x}_{t+1} = f(\mathbf{x}_t), \text{ with } t \in \mathbb{Z}. \tag{1.2}$$

The states of the mappings and of the flows given by difference equations or differential equations, respectively, take values in continuous spaces. However, there are other types of dynamical systems in which their states take values in discrete spaces, such as cellular automata or Boolean differential equations. Figure 1.1 shows four different dynamical systems classify according to discrete/continuous time and discrete/continuous space.

In this book we study dynamical systems of differential equations given by Eq. (1.1). In general, $f : E \to \mathbb{R}^n$ and E is an open subset of $\mathbb{R}^n$.

Definition 1.1. A point $\mathbf{x}^* \in \mathbb{R}^n$ is called an equilibrium point or critical point of (1.6) if $f(\mathbf{x}^*) = 0$.

Flows: $dx/dt = f(x)$ continuous spaces continuous time	discrete time → ← continuous time	Mappings: $x_{t+1} = f(x_t)$ continuous spaces discrete time

| continuous space | discrete space | | continuous space | discrete space |

Boolean differential equation discrete spaces continuous time	discrete time → ← continuous time	Cellular automata discrete space discrete time

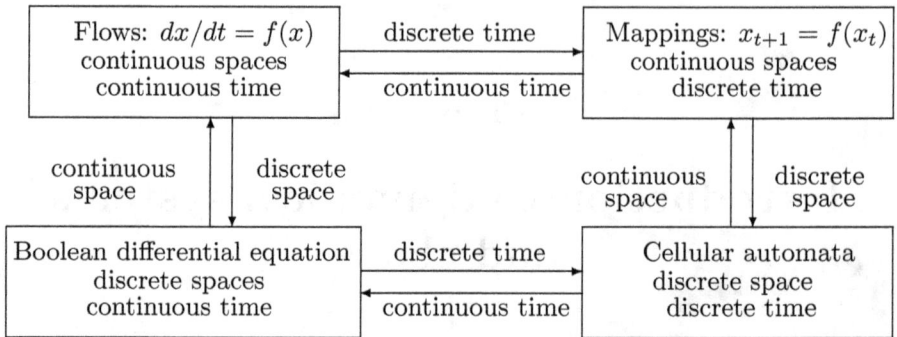

Fig. 1.1   Different types of dynamical systems according to discrete/continuous time and discrete/continuous space.

Equilibria give us important information about the behavior of the continuous dynamical system in the neighborhood of each equilibrium point $\mathbf{x}^*$. Some important theorems have been established based on equilibrium points, for example, one of the most important tools for analyzing the chaotic motion of a nonlinear dynamical system is the well-known Shil'nikov theorem.

Under certain conditions on the function f, the system (1.1) has a unique solution through each point $\mathbf{x}_0 \in E$ defined on a maximal interval of existence $(a, b) \subset \mathbb{R}$. Then the solution of a differential Eq. (1.1) is given by

$$\mathbf{x}(t) = \mathbf{x}(0) + \int_0^t f(s)ds, \tag{1.3}$$

if $f(t)$ is integrable.

Definition 1.2. The phase portrait of a system (1.1) of differential equations with $\mathbf{x} \in \mathbb{R}^n$ is the set of all solution curves of (1.1) in the phase space $\mathbb{R}^n$.

Therefore, the continuous dynamical system defined by (1.1) is the mapping $\phi : \mathbb{R} \times \mathbb{R}^n \to \mathbb{R}^n$ defined by the solution $\mathbf{x}(t, \mathbf{x}_0)$ given by (1.3). The function $f(\mathbf{x})$ on the right-hand side of (1.1) defines a vector field on $\mathbb{R}^n$, i.e., to each point $\mathbf{x} \in \mathbb{R}^n$, the mapping f assigns a vector $f(\mathbf{x})$.

Let $\phi^t : E \to \mathbb{R}^n$ be the flow of the differential Eq. (1.1). If $\phi^t(\mathbf{x}^*) = \mathbf{x}^*$ for all $t \in \mathbb{R}$, then $\mathbf{x}^*$ is an equilibrium point of (1.1) and $\mathbf{x}^*$ is a fixed point of the flow ϕ^t. A fixed point is also called a critical point or a singular point of the vector field $f : E \to \mathbb{R}^n$.

Let $L(\mathbb{R}^n)$ be the linear space of linear operators on $\mathbb{R}^n$. The function $f : \mathbb{R}^n \to \mathbb{R}^n$ is differentiable at $\mathbf{x}_0 \in \mathbb{R}^n$ if there is a linear transformation $f'(\mathbf{x}_0) \in L(\mathbb{R}^n)$ that satisfies

$$\lim_{|h| \to 0} \frac{|f(\mathbf{x}_0 + h) - f(\mathbf{x}_0) - f'(\mathbf{x}_0)h|}{|h|} = 0.$$

Here h is a vector in $\mathbb{R}^n$. If the total derivative exists at $\mathbf{x}_0$, then all the partial derivatives and directional derivatives of f exist at $\mathbf{x}_0$, and for all $\mathbf{x} \in \mathbb{R}^n$, $f'(\mathbf{x}_0)\mathbf{x}$ is the directional derivative of f in the direction $\mathbf{x}$. If we write f using coordinate functions, so that $f = (f_1, f_2, \ldots, f_n)$, then the total derivative can be expressed using the partial derivatives as a matrix. This matrix is called the Jacobian matrix of f at $\mathbf{x}_0$:

$$Jf = \left[\frac{\partial f_i}{\partial x_j} \right].$$

Theorem 1.1. *[Perko (2013)] If $f : \mathbb{R}^n \to \mathbb{R}^n$ is differentiable at $\mathbf{x}_0$, then the partial derivative $\frac{\partial f_i}{\partial x_j}$, $i, j = 1, \ldots, n$, all exist at $\mathbf{x}_0$ and for all $\mathbf{x} \in \mathbb{R}^n$,*

$$f'(\mathbf{x}_0)\mathbf{x} = \sum_{j=1}^{n} \frac{\partial f}{\partial x_j}(\mathbf{x}_0)x_j.$$

A lot of information of a nonlinear system (1.1) is obtained by analyzing the system near its equilibria. The local behavior of a nonlinear system (1.1) near a hyperbolic equilibrium point $\mathbf{x}^*$ is given by the behavior of the linear system

$$\dot{\mathbf{x}} = A\mathbf{x}, \tag{1.4}$$

where $A = Jf(\mathbf{x}^*)$ is called the linearization of (1.1) at $\mathbf{x}^*$. The linear part $A\mathbf{x}$ of f at $\mathbf{x}^*$ gives qualitative information in a neighborhood around the equilibrium point $\mathbf{x}^*$.

Definition 1.3. An equilibrium point $\mathbf{x}^*$ is called hyperbolic equilibrium point of (1.1) if none of the eigenvalues of the Jacobian matrix $Jf(\mathbf{x}^*)$ have zero real part.

The Taylor's theorem is established as follows

Theorem 1.2 (Taylor's Theorem). *Suppose f has $n + 1$ continuous derivatives on an open interval containing $\mathbf{a}$. Then for each $\mathbf{x}$ in the interval,*

$$f(\mathbf{x}) = \left[\sum_{k=0}^{n} \frac{f^{(k)}(\mathbf{a})}{k!}(\mathbf{x} - \mathbf{a})^k \right] + R_{n+1}(\mathbf{x}), \tag{1.5}$$

where the error term $R_{n+1}(\mathbf{x})$ satisfies $R_{n+1}(\mathbf{x}) = \dfrac{f^{(n+1)}(\mathbf{c})}{(n+1)!}(\mathbf{x} - \mathbf{a})^{n+1}$ *for some* $\mathbf{c}$ *between* $\mathbf{a}$ *and* $\mathbf{x}$.

The infinite Taylor series converges to f,

$$f(x) = \sum_{k=0}^{\infty} \frac{f^{(k)}(\mathbf{a})}{k!}(\mathbf{x} - \mathbf{a})^k,$$

if and only if $\lim\limits_{n\to\infty} R_n(\mathbf{x}) = 0$. If $\mathbf{a} = \mathbf{x}^* = \mathbf{0}$ is an equilibrium point of (1.1), then $f(\mathbf{0}) = \mathbf{0}$ and by the Taylor's theorem

$$f(\mathbf{x}) = f'(\mathbf{0})\mathbf{x} + \frac{1}{2}f''(\mathbf{0})(\mathbf{x}, \mathbf{x}) + \ldots$$

The linear part $f'(\mathbf{0})\mathbf{x} = Jf(\mathbf{0})\mathbf{x}$ is a good first approximation to the nonlinear function $f(\mathbf{x})$ near $\mathbf{x}^* = 0$.

1.2 Linear systems

We consider linear maps from $\mathbb{R}^n$ to $\mathbb{R}^n$, which we denoted by $A \in L(\mathbb{R}^n)$. A linear differential equation is given as follows

$$\frac{d\mathbf{x}}{dt} = A\mathbf{x}, \tag{1.6}$$

with $\mathbf{x} \in \mathbb{R}^n$ and $A = (a_{ij})$ an $n \times n$ matrix. Given an initial condition $\mathbf{x}(0) = \mathbf{x}_0 \in \mathbb{R}^n$, the flow of the differential linear Eq. (1.6) is a function $\phi^t(\mathbf{x}_0)$ for $t \in \mathbb{R}$ such that

- $\phi^0(\mathbf{x}_0) = \mathbf{x}_0$,
- $\dfrac{d}{dt}\phi^t(\mathbf{x}_0) = A\phi^t(\mathbf{x}_0)$ for t which it is defined.

Theorem 1.3. *If the $n \times n$ matrix $A = \mathrm{diag}[\lambda_1, \ldots, \lambda_n]$, with $\lambda_i \in \mathbb{R}$ and distinct, then the solution of (1.6) is given by $\mathbf{x}(t) = \mathrm{diag}[e^{\lambda_1 t}, \ldots, e^{\lambda_n t}]\mathbf{x}_0$.*

Proof.

$$\frac{d}{dt}\mathbf{x}(t) = \frac{d}{dt}\mathrm{diag}[e^{\lambda_1 t}, \ldots, e^{\lambda_n t}]\mathbf{x}_0 = \mathrm{diag}[\lambda_1 e^{\lambda_1 t}, \ldots, \lambda_n e^{\lambda_n t}]\mathbf{x}_0 \tag{1.7}$$

$$= \mathrm{diag}[\lambda_1, \ldots, \lambda_n]\mathrm{diag}[e^{\lambda_1 t}, \ldots, e^{\lambda_n t}]\mathbf{x}_0$$

$$= A\,\mathrm{diag}[e^{\lambda_1 t}, \ldots, e^{\lambda_n t}]\mathbf{x}_0 \tag{1.8}$$

$$= A\mathbf{x}(t). \tag{1.9}$$

$\square$

Theorem 1.4. *Let A be an $n \times n$ matrix with real and distinct eigenvalues $\lambda_1, \lambda_2, \ldots, \lambda_n$ and corresponding eigenvector $\{\mathbf{v}_1, \mathbf{v}_2, \ldots, \mathbf{v}_n\}$, then a matrix P always exists and $P^{-1}AP = \mathrm{diag}[\lambda_1, \ldots, \lambda_n]$.*

Proof. The matrix P can always be defined as $P = [\mathbf{v}_1, \mathbf{v}_2, \ldots, \mathbf{v}_n]$. And we know that

$$A\,\mathbf{v}_1 = \lambda_1\,\mathbf{v}_1,$$
$$A\,\mathbf{v}_2 = \lambda_2\,\mathbf{v}_2,$$
$$\vdots$$
$$A\,\mathbf{v}_n = \lambda_n\,\mathbf{v}_n,$$

then

$$AP = P\,\mathrm{diag}[\lambda_1, \ldots, \lambda_n]$$
$$P^{-1}AP = \mathrm{diag}[\lambda_1, \ldots, \lambda_n]. \qquad \square$$

Theorem 1.5. *If the linear operator A of the linear system (1.6) fulfills the Theorem 1.4, then the solution of (1.6) is given by*

$$\mathbf{x}(t) = P\,\mathrm{diag}[\lambda_1, \ldots, \lambda_n]P^{-1}\mathbf{x}(0).$$

Proof. We define the linear transformation of coordinates

$$\mathbf{y} = P^{-1}\mathbf{x},$$

where P is given by the eigenvector of A according to Theorem 1.4. Then

$$\dot{\mathbf{y}} = P^{-1}\dot{\mathbf{x}} = P^{-1}A\mathbf{x}. \qquad (1.10)$$

From the linear transformation we have

$$\mathbf{x} = P\mathbf{y}. \qquad (1.11)$$

If (1.11) is substituted in (1.10), we have

$$\dot{\mathbf{y}} = P^{-1}AP\mathbf{y} = \mathrm{diag}[\lambda_1, \ldots, \lambda_n]\mathbf{y}. \qquad (1.12)$$

According to Theorem 1.3, the solution of (1.12) is

$$\mathbf{y}(t) = \mathrm{diag}[e^{\lambda_1 t}, \ldots, e^{\lambda_n t}]\mathbf{y}(0).$$

From the linear transformation $\mathbf{y}(0) = P^{-1}\mathbf{x}(0)$ and $\mathbf{x}(t) = P\mathbf{y}(t)$. Therefore, the solution of the linear system (1.6) is

$$\mathbf{x}(t) = P\,\mathrm{diag}[e^{\lambda_1 t}, \ldots, e^{\lambda_n t}]P^{-1}\mathbf{x}(0). \qquad \square$$

The above theorems correspond to real and distinct eigenvalues, now, we discuss when eigenvalues are complex. If an eigenvalue $\lambda_j = \alpha_j + i\beta_j$ is complex, then its eigenvector $\mathbf{v}^j = \mathbf{u}^j + i\mathbf{w}^j$ must also be complex. Since A is real, the complex conjugate $\bar{\lambda}_j = \alpha_j - i\beta_j$ is also an eigenvalue and has eigenvector $\bar{\mathbf{v}}^j = \mathbf{u}^j - i\mathbf{w}^j$.

Theorem 1.6. *Let A be an $2n \times 2n$ matrix with complex and distinct eigenvalues $\lambda_j = \alpha_j + i\beta_j$ and $\bar{\lambda}_j = \alpha_j - i\beta_j$ and corresponding complex eigenvectors $\{\mathbf{v}_j = \mathbf{u}_j + i\mathbf{w}_j$ and $\bar{\mathbf{v}}_j = \mathbf{u}_j - i\mathbf{w}_j, j = 1, \ldots, n$, then a matrix P always exists and $P^{-1}AP = \mathrm{diag}[D_1, \ldots, D_n]$, with $D_j = \begin{pmatrix} \alpha_j & -\beta_j \\ \beta_j & \alpha_j \end{pmatrix}$.*

Proof. The matrix P can always be defined as $P = [\mathbf{w}_1, \mathbf{u}_1, \mathbf{w}_2, \mathbf{u}_2, \ldots, \mathbf{w}_n, \mathbf{u}_n]$ because $\mathbf{w}_1, \mathbf{u}_1, \mathbf{w}_2, \mathbf{u}_2, \ldots, \mathbf{w}_n, \mathbf{u}_n$ is a basis for $\mathbb{R}^{2n}$. And we have that

$$A\,\mathbf{v}_j = \lambda_j\,\mathbf{v}_j,$$
$$A\,(\mathbf{u}_j + i\mathbf{w}_j) = (\alpha_j + i\beta_j)\,(\mathbf{u}_j + i\mathbf{w}_j),$$
$$A\,\mathbf{u}_j + iA\,\mathbf{w}_j = \alpha_j\mathbf{u}_j - \beta_j\mathbf{w}_j + i(\alpha_j\,\mathbf{w}_j + \beta_j\,\mathbf{u}_j),$$

equating the real and imaginary parts yields

$$A\,\mathbf{u}_j = \alpha_j\mathbf{u}_j - \beta_j\mathbf{w}_j$$
$$A\,\mathbf{w}_j = \beta_j\,\mathbf{u}_j + \alpha_j\,\mathbf{w}_j,$$

then

$$\begin{aligned} AP &= [A\mathbf{w}_1, A\mathbf{u}_1, A\mathbf{w}_2, A\mathbf{u}_2, \ldots, A\mathbf{w}_n, A\mathbf{u}_n] \\ &= [\alpha_1\mathbf{w}_1 + \beta_1\mathbf{u}_1, -\beta_1\mathbf{w}_1 + \alpha_1\mathbf{u}_1, \alpha_2\mathbf{w}_2 + \beta_2\mathbf{u}_2, \\ &\quad\ - \beta_2\mathbf{w}_2 + \alpha_2\mathbf{u}_2, \ldots, \alpha_n\mathbf{w}_n + \beta_n\mathbf{u}_n, -\beta_n\mathbf{w}_n + \alpha_n\mathbf{u}_n] \\ &= [\mathbf{w}_1, \mathbf{u}_1, \mathbf{w}_2, \mathbf{u}_2, \ldots, \mathbf{w}_n, \mathbf{u}_n]\,\mathrm{diag}[D_1, \ldots, D_n] \\ &= P\,\mathrm{diag}[D_1, \ldots, D_n]. \end{aligned}$$

Therefore

$$P^{-1}AP = \mathrm{diag}[D_1, D_2, \ldots, D_n]. \qquad \square$$

If the $2n \times 2n$ real matrix A fulfills the Theorem 1.6, then the solution of the initial value problem

$$\dot{\mathbf{x}} = A\mathbf{x}, \tag{1.13}$$
$$\mathbf{x}(0) = \mathbf{x}_0, \tag{1.14}$$

is given by

$$\mathbf{x}(t) = P\,\mathrm{diag}\,e^{\alpha_j t}\begin{pmatrix} \cos\beta_j t & -\sin\beta_j t \\ \sin\beta_j t & \cos\beta_j t \end{pmatrix} P^{-1}\mathbf{x}_0. \tag{1.15}$$

Each matrix $R_j = \begin{pmatrix} \cos \beta_j t & -\sin \beta_j t \\ \sin \beta_j t & \cos \beta_j t \end{pmatrix}$ represents a rotation through βt radians.

Theorem 1.7. *If the $2n - k \times 2n - k$ real matrix A has k distinct real eigenvalues λ_j and corresponding eigenvectors $\mathbf{v}_j$, $j = 1, \ldots, k$, and $2(n-k)$ distinct complex eigenvalues $\lambda_j = \alpha + i\beta$ and $\bar{\lambda}_j = \alpha - i\beta$ and corresponding eigenvectors $\mathbf{v}_j = \mathbf{u} + i\mathbf{w}$ and $\mathbf{v}_j = \mathbf{u} - i\mathbf{w}$, $j = k+1, \ldots, n$, then the matrix*

$$P = [\mathbf{v}_1, \ldots, \mathbf{v}_k, \mathbf{w}_{k+1}, \mathbf{u}_{k+1}, \ldots, \mathbf{w}_n, \mathbf{u}_n]$$

is invertible and

$$P^{-1}AP = \mathrm{diag}[\lambda_1, \ldots, \lambda_k, D_{k+1}, \ldots, D_n],$$

where $D_j = \begin{pmatrix} \alpha_j & -\beta_j \\ \beta_j & \alpha_j \end{pmatrix}$.

Proof. The proof is similar to the proofs of Theorems 1.4 and 1.6. □

Theorem 1.8 (The Fundamental Theorem for Linear Systems). *[Perko (2013)] Let A be an $n \times n$ matrix. Then for a given $\mathbf{x}_0 \in \mathbb{R}^n$, the initial value problem*

$$\dot{\mathbf{x}} = A\mathbf{x},$$
$$\mathbf{x}(0) = \mathbf{x}_0,$$

has a unique solution given by

$$\mathbf{x}(t) = e^{At}\mathbf{x}_0.$$

Proof. The proof can be seen in [Perko (2013)]. □

The mapping $\phi^t = e^{At}$ satisfies the following basic properties for all $\mathbf{x} \in \mathbb{R}^n$:

- $\phi^0(\mathbf{x}) = \mathbf{x}$.
- $\phi^s(\phi^t(\mathbf{x})) = \phi^{s+t}(\mathbf{x})$ for all $s, t \in \mathbb{R}$.
- $\phi^{-t}(\phi^t(\mathbf{x})) = \phi^t(\phi^{-t}(\mathbf{x})) = \mathbf{x}$ for all $t \in \mathbb{R}$.

Definition 1.4. A fundamental matrix solution of a linear system $\dot{\mathbf{x}} = A\mathbf{x}$ is any nonsingular $n \times n$ matrix function $\Phi(t)$ that fulfills $\Phi'(t) = A\Phi(t)$ for all $t \in \mathbb{R}$

A nonhomogeneous linear system is given by

$$\dot{\mathbf{x}} = A\mathbf{x} + \mathbf{b}(t), \qquad (1.16)$$

where A is an $n \times n$ matrix and $\mathbf{b}(t)$ is a continuous vector valued function. Once we have found the solution of the linear system, it is easy to find the solution of the affine linear system as follows

Theorem 1.9. *[Perko (2013)] If $\Phi(t)$ is any fundamental matrix solution of a linear system, then the solution of the nonhomogeneous linear system (1.17) and the initial condition $\mathbf{x}(0) = \mathbf{x}_0$ is unique and is given by*

$$\mathbf{x}(t) = \Phi(t)\Phi^{-1}(0)\mathbf{x}_0 + \int_0^t \Phi(t)\Phi^{-1}(\tau)b(\tau)d\tau.$$

Proof. The proof can be seen in [Perko (2013)]. □

The affine linear systems are given when $b(t)$ is constant, this kind of systems can be transformed to a linear system.

Theorem 1.10. *Let A be an $n \times n$ nonsingular matrix and $\mathbf{b}$ is a constant vector in $\mathbb{R}^n$. Then the affine linear system*

$$\dot{\mathbf{x}} = A\mathbf{x} + \mathbf{b}, \qquad (1.17)$$

can be transformed to a linear system

$$\dot{\mathbf{y}} = A\mathbf{y}.$$

Proof. Because A is nonsingular, then it is possible to compute the equilibrium point of the affine linear system which is given by

$$\mathbf{x}^* = -A^{-1}\mathbf{b}.$$

Now, we consider the following variable change $\mathbf{x} = \mathbf{y} + \mathbf{x}^*$. We have $\dot{\mathbf{x}} = \dot{\mathbf{y}}$, then the system (1.17) can be given in terms of $\mathbf{y}$ state vector as follows

$$\dot{\mathbf{y}} = A(\mathbf{y} + \mathbf{x}^*) + \mathbf{b}$$
$$\dot{\mathbf{y}} = A\mathbf{y} + A\mathbf{x}^* + \mathbf{b}$$
$$\dot{\mathbf{y}} = A\mathbf{y} + A(-A^{-1}\mathbf{b}) + \mathbf{b}$$
$$\dot{\mathbf{y}} = A\mathbf{y}. \qquad \square$$

Definition 1.5. The flow of the linear system is the set of mappings e^{At} : $\mathbb{R}^n \to \mathbb{R}^n$ that describes the motion of points $\mathbf{x}_0 \in \mathbb{R}^n$ along trajectories of the linear system.

Definition 1.6. A hyperbolic flow is the flow $e^{At} : \mathbb{R}^n \to \mathbb{R}^n$ such that all eigenvalues of the $n \times n$ matrix A have nonzero real part. A hyperbolic linear system is a linear system with hyperbolic flow.

Definition 1.7. An invariant subspace $E \subset \mathbb{R}^n$ with respect to the flow $e^{At} : \mathbb{R}^n \to \mathbb{R}^n$ fulfills $e^{At}E \subset E$ for all $t \in \mathbb{R}$.

1.3 Stability theory

In this section we study three subspaces that are determined by the eigenvalues of a linear operator of a linear system

$$\dot{\mathbf{x}} = A\mathbf{x}. \tag{1.18}$$

We start by considering a $n \times n$ real matrix A with distinct real eigenvalues and suppose that A has k negative eigenvalues $\lambda_1, \ldots, \lambda_k$ and $n - k$ positive eigenvalues $\lambda_{k+1}, \ldots, \lambda_n$ and the set of corresponding eigenvectors $\{\mathbf{v}_1, \ldots, \mathbf{v}_n\}$. Then two subspaces of the linear system (1.18) are defined as follows

- A stable subspace E^s is the subspace spanned by $\{\mathbf{v}_1, \ldots, \mathbf{v}_k\}$, i.e.,

$$E^s = \text{Span}\{\mathbf{v}_1, \ldots, \mathbf{v}_k\}.$$

- An unstable subspace E^u is the subspace spanned by $\{\mathbf{v}_{k+1}, \ldots, \mathbf{v}_n\}$, i.e.,

$$E^u = \text{Span}\{\mathbf{v}_{k+1}, \ldots, \mathbf{v}_n\}.$$

Each negative or positive eigenvalue and its corresponding eigenvector spans a one dimension stable or unstable subspace, respectively. Every initial condition $\mathbf{x}_0 \in E^s$ generates a trajectory $\phi(\mathbf{x}_0)$ that tends to the equilibrium point and every initial condition $\mathbf{x}_0 \in E^u$ generates a trajectory $\phi(\mathbf{x}_0)$ that gets away from the equilibrium point.

Theorem 1.11. *Let E^s and E^u be the stable and unstable subspaces of a linear system (1.18), respectively. Then these subspaces are invariant with respect to the flow $e^{At} : \mathbb{R}^n \to \mathbb{R}^n$.*

Proof. We need to prove that $e^{At}E^s \subset E^s$ and $e^{At}E^u \subset E^u$ for all $t \in \mathbb{R}$. If $\mathbf{v} \in E^s$ then it can be given as

$$\mathbf{v} = \sum_{j=1}^{k} c_j \mathbf{v}_j,$$

with $c_j \in \mathbb{R}$ and $j = 1, \ldots, k$. Then

$$e^{At}\mathbf{v} = \left(I + At + A^2t^2/2! + \ldots\right)\mathbf{v}$$

$$I\mathbf{v} = \sum_{j=1}^{k} c_j\mathbf{v_j},$$

$$A t\mathbf{v} = \sum_{j=1}^{k} \lambda_j t c_j\mathbf{v_j},$$

$$(A^2t^2/2!)\mathbf{v} = \sum_{j=1}^{k} (\lambda_j^2 t^2/2!) c_j\mathbf{v_j},$$

$$\vdots \qquad\qquad \vdots$$

$$e^{At}\mathbf{v} = \sum_{j=1}^{k} e^{\lambda_j t} c_j\mathbf{v_j} \subset E^s,$$

for all $t \in \mathbb{R}$. Therefore

$$e^{At} E^s \subset E^s.$$

If $\mathbf{v} \in E^u$, then

$$\mathbf{v} = \sum_{j=k+1}^{n} c_j\mathbf{v_j},$$

where $c_j \in \mathbb{R}$ and $j = k+1, \ldots, n$.

$$e^{At}\mathbf{v} = \left(I + At + A^2t^2/2! + \ldots\right)\mathbf{v}$$

$$e^{At}\mathbf{v} = \sum_{j=k+1}^{n} e^{\lambda_j t} c_j\mathbf{v_j} \subset E^u,$$

for all $t \in \mathbb{R}$. Therefore

$$e^{At} E^u \subset E^u. \qquad\qquad \square$$

Now we consider the case of complex eigenvalues of a real matrix A. A complex eigenvalue comes in pair with its complex conjugate eigenvalue. The real part of a complex eigenvalue determines the kind of stability of a subspace in the same way than real eigenvalues. Let $\mathbf{v}_j = \mathbf{u}_j + i\mathbf{w}_j$ be an eigenvector of the real matrix A corresponding to an eigenvalue $\lambda_j = \alpha_j + i\beta_j$. Then

- A stable subspace E^s is the subspace spanned by
$$E^s = \text{Span}\{\mathbf{u}_j, \mathbf{w}_j | \alpha_j < 0\}.$$

- An unstable subspace E^u is the subspace spanned by
$$E^u = \text{Span}\{\mathbf{u}_j, \mathbf{w}_j | \alpha_j > 0\}.$$

If the matrix A has an eigenvalue with only pure imaginary part, $\alpha_j = 0$ and $\beta_j \neq 0$. Then

- A center subspace E^c is the subspace spanned by
$$E^c = \text{Span}\{\mathbf{u}_j, \mathbf{w}_j | \alpha_j = 0\}.$$

These threes subspaces of $\mathbb{R}^n$, E^s, E^u and E^c, are spanned by the eigenvectors $\mathbf{v}_j = \mathbf{u}_j + i\mathbf{w}_j$. Because $\mathbf{u}_j$ and $\mathbf{w}_j$ corresponding to a complex eigenvalue then these eigenvectors span a two dimension subspace. The stability of the subspace is given by the real part of the complex eigenvalue.

The set of all eigenvector $B = \{\mathbf{v}_1, \ldots, \mathbf{v}_k, \mathbf{u}_{k+1}, \mathbf{w}_{k+1}, \ldots, \mathbf{u}_m, \mathbf{w}_m\}$ is a basis of $\mathbb{R}^n$, then $\mathbb{R}^n = E^s \oplus E^u \oplus E^c$. Furthermore, E^s, E^u and E^c are invariant with respect to the flow e^{At} of (1.18).

There are different topologies that are given by stable, unstable and center subspaces. We start by considering a linear system (1.18) in $\mathbb{R}^2$ which is given by a 2×2 real matrix A, with $\mathbf{x} = (x_1, x_2)^T \in \mathbb{R}^2$. The characteristic polynomial of A is $p(\lambda) = \lambda^2 - \tau\lambda + \delta$ and its discriminant $\Delta = \tau^2 - 4\delta$, where $\delta = \det A$, $\tau = \text{trace } A$. The classification of phase portrait of two type of linear operator $A \in \mathbb{R}^{2 \times 2}$ is given. The former case considers real eigenvalues of the linear operator as follows

$$A = \begin{bmatrix} \lambda_1 & 0 \\ 0 & \lambda_2 \end{bmatrix}. \tag{1.19}$$

It follows from the fundamental theorem of linear systems that the solution of the initial value problem with $\mathbf{x}(0) = \mathbf{x}_0$ is respectively given by

$$\mathbf{x}(t) = \begin{bmatrix} e^{\lambda_1 t} & 0 \\ 0 & e^{\lambda_2 t} \end{bmatrix} \mathbf{x}_0. \tag{1.20}$$

The phase portrait of a linear system (1.18) by applying the linear operator A with real eigenvalues results from the above solution in a stable node, unstable node or saddle at the origin.

- A stable node is given for $\lambda_1 \leq \lambda_2 < 0$. $\delta = \lambda_1\lambda_2 > 0$, $\tau = \lambda_1 + \lambda_2 < 0$ and $\Delta = (\lambda_1 - \lambda_2)^2 > 0$. If $\lambda_1 = \lambda_2$ then the origin is called a proper node. If $\lambda_1 < \lambda_2$ then the origin is called an improper node.

- An unstable node is given for $\lambda_1 \geq \lambda_2 > 0$. $\delta = \lambda_1 \lambda_2 > 0$, $\tau = \lambda_1 + \lambda_2 > 0$ and $\Delta = (\lambda_1 - \lambda_2)^2 > 0$. If $\lambda_1 = \lambda_2$ then the origin is called a proper node. If $\lambda_1 > \lambda_2$ then the origin is called an improper node.
- A saddle is given for $\lambda_1 < 0 < \lambda_2$. $\delta_1 = \lambda_1 \lambda_2 < 0$ and $\Delta = (\lambda_1 - \lambda_2)^2 > 0$. The four non-zero trajectories or solution curves that approach the equilibrium point at the origin as $t \to \pm\infty$ are called separatrices of the system.

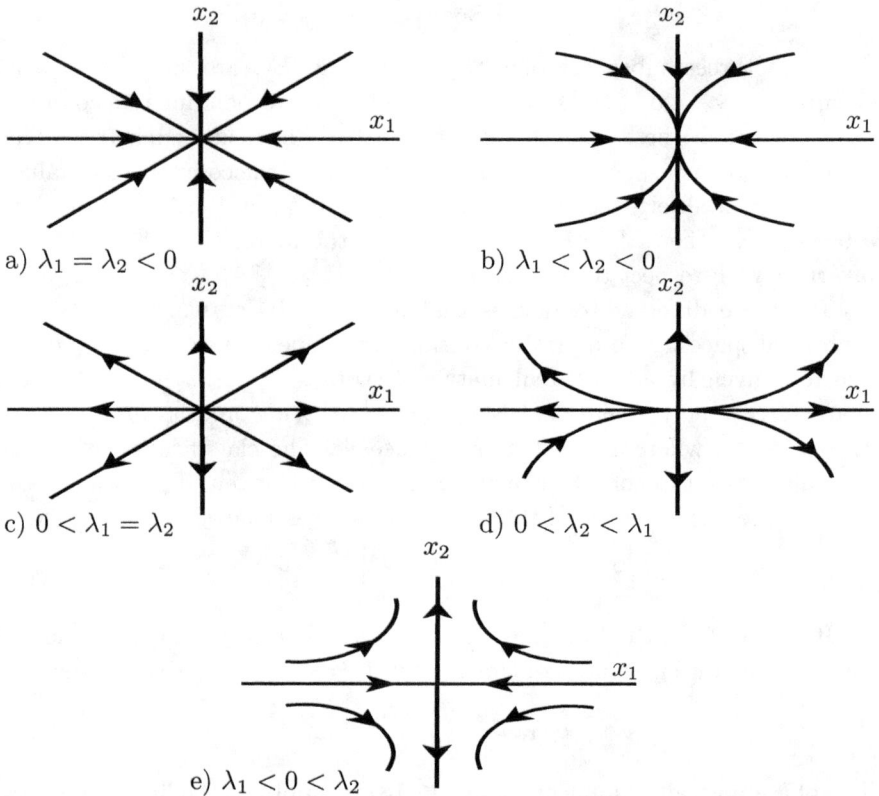

Fig. 1.2   Different phase portraits of dynamical systems (1.18). Stable node at the origin: (a) proper and (b) improper. Unstable node at the origin: (c) proper and (d) improper. (e) A saddle at the origin.

From the three previous cases we observe that if $\delta < 0$ then (1.18) has a saddle at the origin. If $\delta > 0$ and $\Delta \geq 0$ then (1.18) has a node at the origin. Figure 1.2 shows different phase portraits of dynamical systems (1.18).

Figure 1.2(a) Proper stable node at the origin. Figure 1.2(b) Improper stable node at the origin. Figure 1.2(c) Proper unstable node at the origin. Figure 1.2(d) Improper unstable node at the origin. Figure 1.2(e) A saddle at the origin.

Now, it is considered the phase portrait of a linear system (1.18) by applying the linear operator $A \in \mathbb{R}^{2 \times 2}$ that has complex eigenvalues as follows

$$A = \begin{bmatrix} \alpha & -\beta \\ \beta & \alpha \end{bmatrix}, \qquad (1.21)$$

with $\alpha, \beta \in \mathbb{R}$ and $\beta \neq 0$. The solution of the initial value problem with $\mathbf{x}(0) = \mathbf{x}_0$ is given by

$$\mathbf{x}(t) = e^{\alpha t} \begin{bmatrix} \cos \beta t & -\sin \beta t \\ \sin \beta t & \cos \beta t \end{bmatrix} \mathbf{x}_0. \qquad (1.22)$$

The phase portrait of a linear system (1.18) by applying the linear operator A results in a focus at the origin.

- A stable focus at the origin is given for $\alpha < 0$. $\delta = \alpha^2 + \beta^2 > 0$, $\tau = 2\alpha < 0$ and $\Delta = -4\beta^2 < 0$. If $\beta > 0$ then the oscillation is clockwise. If $\beta < 0$ then the oscillation is counterclockwise.
- An unstable focus at the origin is given for $\alpha > 0$. $\delta = \alpha^2 + \beta^2 > 0$, $\tau = 2\alpha > 0$ and $\Delta = -4\beta^2 < 0$. If $\beta > 0$ then the oscillation is clockwise. If $\beta < 0$ then the oscillation is counterclockwise.

From the two previous cases we observe that $\delta > 0$ and $\Delta < 0$. Figure 1.3 shows stable focus at the origin: (a) $\beta > 0$ (b) $\beta < 0$, and unstable focus at the origin: (c) $\beta < 0$ and (d) $\beta > 0$.

Definition 1.8. A stable node or focus of (1.18) is called a sink of the linear system and an unstable node or focus of (1.18) is called a source of the linear system.

If $\alpha = 0$, the phase portrait of a linear system (1.18) by applying the linear operator A_2.

- A center at the origin is given for $\alpha = 0$. $\delta = \beta^2 > 0$, $\tau = 0$ and $\Delta = -4\beta^2 < 0$. If $\beta > 0$ then the oscillation is clockwise. If $\beta < 0$ then the oscillation is counterclockwise.

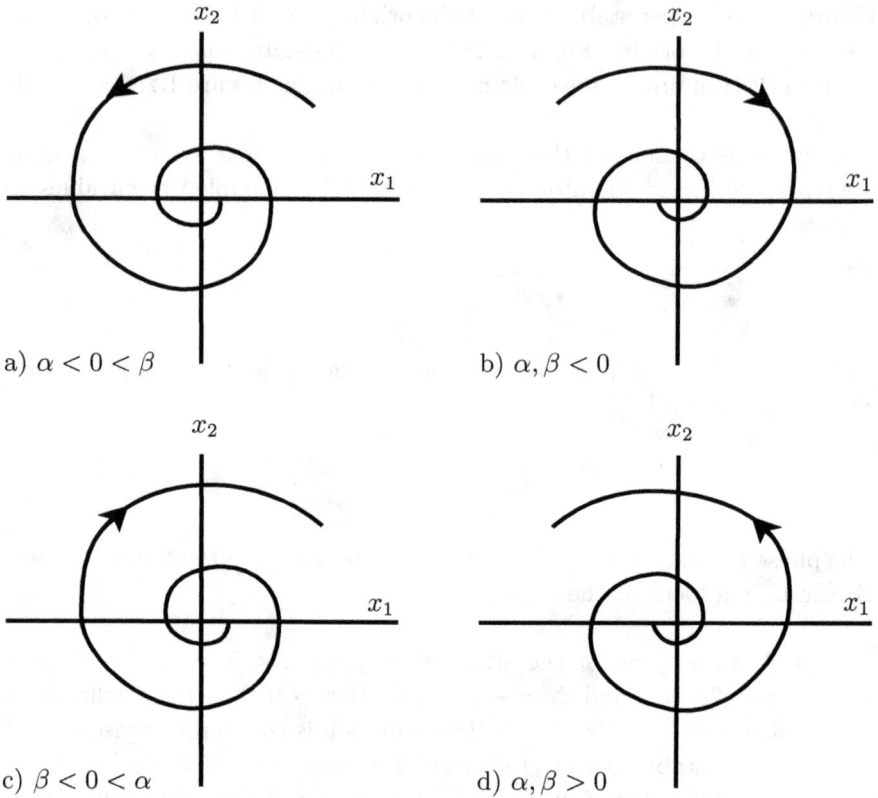

a) $\alpha < 0 < \beta$

b) $\alpha, \beta < 0$

c) $\beta < 0 < \alpha$

d) $\alpha, \beta > 0$

Fig. 1.3   Different phase portraits of dynamical systems (1.18). Stable focus at the origin: (a) $\beta > 0$ (b) $\beta < 0$. Unstable focus at the origin: (c) $\beta < 0$ and (d) $\beta > 0$.

From the previous case we observe that a center occurs if $\delta > 0$ and $\Delta < 0$. Figure 1.4 shows a center at the origin: (a) $\beta > 0$ (b) $\beta < 0$. The above results support the following theorem.

Theorem 1.12. *An equilibrium point at the origin of a linear system* $\dot{\mathbf{x}} = A\mathbf{x}$ *is characterized as follows*

- *If $\delta < 0$ then the system has a saddle at the origin.*
- *If $\delta > 0$ and $\Delta \geq 0$ then the system has a node at the origin; it is stable if $\tau < 0$ and unstable if $\tau > 0$.*
- *If $\delta > 0$ and $\Delta < 0$ then the system has a focus at the origin; it is stable if $\tau < 0$ and unstable if $\tau > 0$.*
- *If $\delta > 0$ and $\tau = 0$ then the system has a center at the origin.*

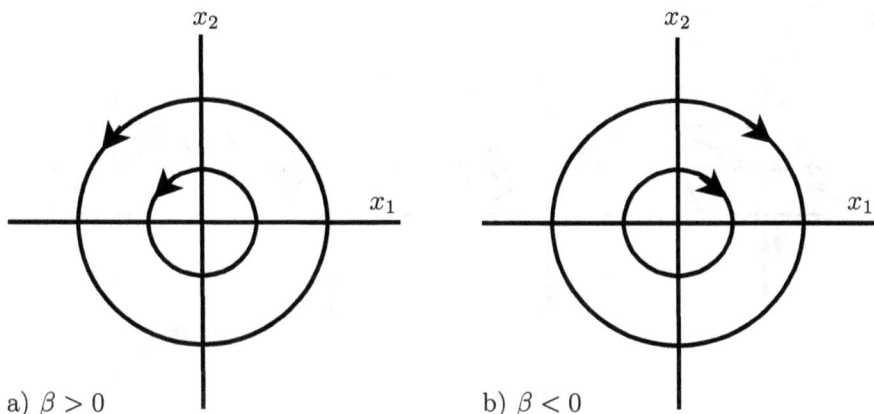

a) $\beta > 0$

b) $\beta < 0$

Fig. 1.4   Different phase portraits of dynamical systems (1.18). A center at the origin: (a) $\beta > 0$ (b) $\beta < 0$.

Now we discuss the phase portrait in the Three-Dimensional Space. The real matrix of a three-dimensional linear system has 3 eigenvalues: λ_1, λ_2 and λ_3, one of them must be real and the other two can be either both real or complex-conjugate. Depending on the types and signs of the eigenvalues, there are the following cases by a hyperbolic equilibrium at the origin.

- Node when all eigenvalues are real $\lambda_i \in \mathbb{R}$, $j = 1, 2, 3$, and have the same sign. If $\lambda_j < 0$, then the node is stable. If $\lambda_j > 0$, then the node is unstable.
- Saddle when all eigenvalues are real $\lambda_i \in \mathbb{R}$ and at least one of them is positive and at least one is negative.
- Focus-Node when it has one real eigenvalue $\lambda_1 \in \mathbb{R}$ and a pair of complex-conjugate eigenvalues $\lambda_{2,3} \in \mathbb{C}$, and all eigenvalues have real parts of the same sign; The equilibrium is stable (unstable) when the sign is negative (positive).
- Saddle-Focus when it has one real eigenvalue $\lambda_1 \in \mathbb{R}$ with the sign opposite to the sign of the real part of a pair of complex-conjugate eigenvalues $\lambda_{2,3} \in \mathbb{C}$.

Figure 1.5 shows different phase portraits of the dynamical system (1.18) in the space $\mathbb{R}^3$. Figure 1.5(a) shows a stable node where $\lambda_{1,2,3} \in \mathbb{R}^-$. Figure 1.5(b) shows a unstable node where $\lambda_{1,2,3} \in \mathbb{R}^+$. Figure 1.5(c) shows a saddle where $\lambda_{1,2} \in \mathbb{R}^-$ and $\lambda_3 \in \mathbb{R}^+$. Figure 1.5(d) shows a saddle where $\lambda_1 \in \mathbb{R}^-$ and $\lambda_{2,3} \in \mathbb{R}^+$. Figure 1.5(e) shows a stable focus-node where $\lambda_1 \in \mathbb{R}^-$ and $\lambda_{2,3} \in \mathbb{C}^-$. Figure 1.5(f) shows a unstable focus-node

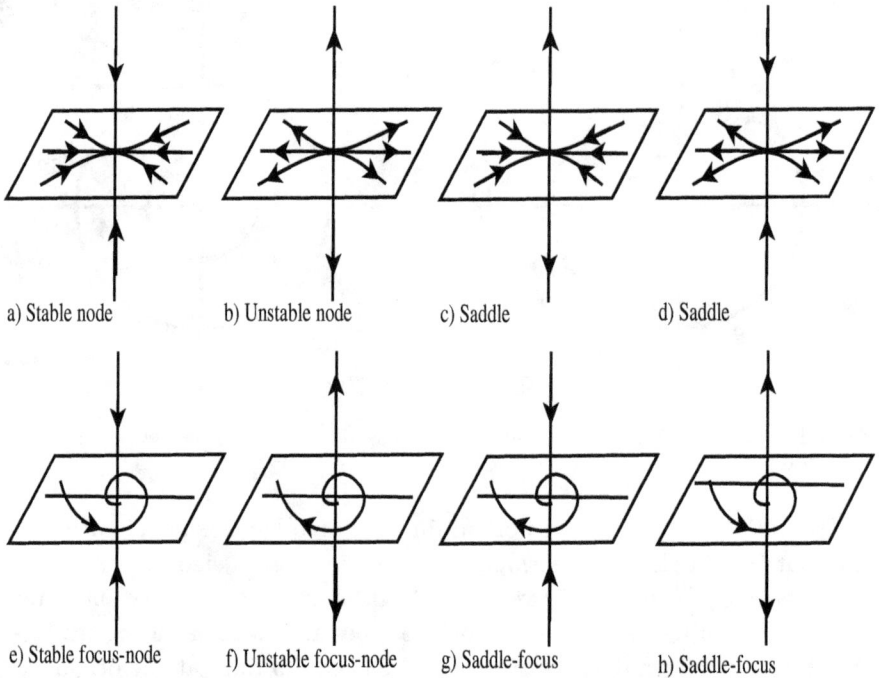

a) Stable node   b) Unstable node   c) Saddle   d) Saddle

e) Stable focus-node   f) Unstable focus-node   g) Saddle-focus   h) Saddle-focus

Fig. 1.5   Different phase portraits of dynamical systems (1.18).   (a) Stable node. (b) Unstable node. (c) Saddled (d) Saddle. (e) Stable focus-node. (f) Unstable focus-node. (g) Saddle-focus. (h) Saddle-focus.

where $\lambda_1 \in \mathbb{R}^+$ and $\lambda_{2,3} \in \mathbb{C}^+$. Figure 1.5(g) shows a saddle-focus where $\lambda_1 \in \mathbb{R}^-$ and $\lambda_{2,3} \in \mathbb{C}^+$. Fig. 1.5 h) shows a saddle-focus where $\lambda_1 \in \mathbb{R}^+$ and $\lambda_{2,3} \in \mathbb{C}^-$. Note that saddle equilibria are always unstable.

Definition 1.9. A saddle-node equilibrium point x^* of (1.18), is called a type-k saddle-node equilibrium point if A has k eigenvalues with positive real part and $n - k$ with negative real part.

1.4   Lyapunov exponents

In this section, Lyapunov exponent is studied, named after the Russian mathematician A. M. Lyapunov. A Lyapunov exponent is a quantity that indicates the convergence or divergence of trajectories of a dynamical system if it is positive or negative, respectively. A positive Lyapunov exponent is related to the sensitive dependence on the initial condition of the solution of a dynamical system.

$$\frac{d\mathbf{x}}{dt} = f(\mathbf{x}), \text{ with } \mathbf{x} \in E \subset \mathbb{R}^n. \tag{1.23}$$

A dynamical system like (1.23) has the same number of Lyapunov exponents as the dimension of the phase space. Therefore, the rate of convergence or divergence can be different for different orientations of initial separation vector.

Definition 1.10. The spectrum of Lyapunov exponents is composed of Lyapunov exponents equal in number to the dimensionality of the phase space.

It is common to refer to the largest one as the maximal Lyapunov exponent (MLE), because it determines a notion of predictability for a dynamical system.

Lyapunov exponents measure the growth rates of generic perturbations, in a regime where their evolution is ruled by linear equations,

$$\frac{d\mathbf{x}}{dt} = J(t)\mathbf{x}$$

where $\mathbf{x}$ is an n-dimensional vector and J is the Jacobian of a suitable velocity field f in the context of deterministic dynamical systems, computed along a trajectory $\mathbf{x}(t)$ that satisfies the ordinary differential Eq. (1.23).

If $\mathbf{x}(t) = \mathbf{x}_0$ is a solution (i.e. if $f(\mathbf{x}_0) = 0$), then, the stability of this fixed point is quantified by the eigenvalues of the (constant) operator J. In this simple case, the LEs λ_i are the real parts of the eigenvalues.

The Jacobian J defines the evolution of the tangent vectors, given by the matrix M, via the equation

$$\dot{M} = JM.$$

The matrix M describes how a small change at the point $\mathbf{x}(0)$ propagates to the final point $\mathbf{x}(t)$. The limit

$$\Lambda = \lim_{t\to\infty} \frac{1}{2t} \log(M(t)M^T(t))$$

defines a matrix Λ. The Lyapunov exponents λ_i are defined by the eigenvalues of Λ, i.e., the finite-time LE is defined from the knowledge of the eigenvalue α_i of $M(t)M^T(t)$ as follows

$$\lambda_i(t) = \frac{\ln \alpha_i(t)}{2t}.$$

It is necessary to consider the infinite time limit, to determine the asymptotic behaviour. This leads to the following definition of LE,

$$\lambda_i = \lim_{t \to \infty} \sup \lambda_i(t).$$

The conditions for the existence of the limit are given by the Oseledets theorem [Oseledec (1968)].

The sum of the Lyapunov exponents is a measure of the rate of volume contraction or expansion of a solution in phase space.

Definition 1.11. A conservative system is given if $\sum_{j=1}^{n} \hat{\lambda}_j = 0$ the volume of a solution in phase space is conserved. A dissipative system is given if $\sum_{j=1}^{n} \hat{\lambda}_j < 0$ the volume of a solution in phase space is contracted. A expanding system is given if $\sum_{j=1}^{n} \hat{\lambda}_j > 0$ the volume of a solution in phase space is expanding.

Note that a dynamical system has an attractor only when

$$\sum_{j=1}^{n} \hat{\lambda}_j \leq 0.$$

In particular from the knowledge of the Lyapunov spectrum it is possible to obtain the so-called Lyapunov dimension (or Kaplan–Yorke dimension) D_{KY}, which is defined as follows:

$$D_{KY} = k + \sum_{i=1}^{k} \frac{\lambda_i}{|\lambda_{k+1}|}$$

where k is the maximum integer such that the sum of the k largest exponents is still non-negative. D_{KY} represents an upper bound for the information dimension of the system [Kaplan and Yorke (1979)].

1.5   Chaos

Lyapunov exponents has been used to identify different types of attractors on dynamical systems. For example, an attractor is only possible for a one-dimensional dynamical system if $\hat{\lambda}_1 < 0$, then the attractor is an equilibrium point in the phase space and the system is dissipative. For a two-dimensional dynamical system there are two Lyapunov exponents: $\hat{\lambda}_1$ and $\hat{\lambda}_2$. The system is dissipative if $\hat{\lambda}_1 + \hat{\lambda}_2 < 0$, then there exist the following attractors: $(\hat{\lambda}_1, \hat{\lambda}_2) = (-, -)$ indicates that all trajectories converge to a point; $(\hat{\lambda}_1, \hat{\lambda}_2) = (-, 0)$ indicates that the trajectory oscillates in a limit cycle in the phase space.

For a three-dimensional dynamical system there are four options that generate an attractor. Therefore, the system is dissipative if $\hat{\lambda}_1 + \hat{\lambda}_2 + \hat{\lambda}_3 < 0$. If $(\hat{\lambda}_1, \hat{\lambda}_2, \hat{\lambda}_3) = (-, -, -)$ indicates that all trajectories converge to a point, then we have a point-like attractor, $(-, -, 0)$ the attractor is a limit cycle, $(-, 0, 0)$ torus with two-frequency quasi-periodicity, and $(-, 0, +)$ strange chaotic attractor. There are several definitions of chaos and one of them is based on the Lyapunov exponents.

Another definition of chaos has been given based on homoclinic and heteroclinic cycles that play an important role in the study of chaotic dynamics. The famous Shil'nikov theorems for differential vector fields showed that the existence of a homoclinic or a heteroclinic cycle implies the existence of a countable number of horseshoes in a neighborhood of this cycle under some conditions [Wiggins *et al.* (2003); Chua *et al.* (2001); Wiggins (2013); Tresser (1984)].

Definition 1.12. A homoclinic orbit is a trajectory $\mathbf{x}(t)$ that connects an equilibrium $\mathbf{x}^*$ to itself, i.e., $\mathbf{x}(t) \rightarrow \mathbf{x}^*$ as $t to \pm \infty$. On the other hand, a heteroclinic orbit is a trajectory $\mathbf{x}(t)$ that connects two different equilibria $\mathbf{x}_1^*$ and $\mathbf{x}_2^*$, i.e., The heteroclinic orbit from $\mathbf{x}_1^*$ to $\mathbf{x}_2^*$ is given when $\mathbf{x}_1(t) \rightarrow \mathbf{x}_2^*$ as $t \rightarrow +\infty$ and $\mathbf{x}_1(t) \rightarrow \mathbf{x}_1^*$ as $t \rightarrow -\infty$. The another heteroclinic orbit from $\mathbf{x}_2^*$ to $\mathbf{x}_1^*$ is given when $\mathbf{x}_2(t) \rightarrow \mathbf{x}_1^*$ as $t \rightarrow +\infty$ and $\mathbf{x}_2(t) \rightarrow \mathbf{x}_2^*$ as $t \rightarrow -\infty$. The heteroclinic cycle is given by $\mathbf{x}_1(t)$ and $\mathbf{x}_2(t)$.

The Shil'nikov Theorems about homoclinic and heteroclinic chaos can be consulted in the reference [Silva (1993)] and are given as follows:

Theorem 1.13. *Given the third order autonomous system $dx/dt = f(x)$, where f is a C^2 vector field on $\mathbb{R}^3$. Let x^* be a equilibrium point. Suppose: (i) The equilibrium point is a saddle focus whose characteristic eigenvalues γ, $\sigma \pm iw$, with γ, σ, $w \in \mathbb{R}$ satisfy the Shil'nikov inequality $|\gamma| > |\sigma| > 0$, and (ii) There exists a homoclinic orbit H based at x^*. Then: (i) the Shil'nikov map defined in a neighborhood of H possesses a countable number of Smale horseshoes in its discrete dynamics. (ii) For any sufficiently small C^1-perturbation of f, the perturbed system has at least a finite number of*

Smale horseshoes in the discrete dynamics of the Shil'nikov map defined near of H (homoclinic chaos).

Theorem 1.14. *Given the conditions of Theorem 1.13. Let x_1^* and x_2^* be two distinct equilibrium points. Suppose: (i) both equilibria are saddle foci that satisfy the Shil'nikov inequality with the further constraint $\gamma_1 \gamma_2 > 0$ and $\sigma_1 \sigma_2 > 0$, and (ii) There is a heteroclinic loop H joining x_1^* to x_2^*. Then, the Shil'nikov map defines on a neighborhood of H possesses a countable number of Smale horseshoes in its discrete dynamics (heteroclinic chaos).*

Chapter 2

Systems with self-excited attractors

There are two types of vector fields, those with equilibrium points and those without equilibrium points. If vector fields with equilibrium points are considered, there are two classes of attractors according to [Leonov *et al.* (2011a)], which are defined as follows: the first class is given by those classical attractors excited from unstable equilibria called self-excited attractors whose basin of attraction intersects at least a neighborhood of an equilibrium point [Dudkowski *et al.* (2016)], and the second class is called hidden attractors whose basin of attraction does not contain neighborhoods of equilibria. If vector fields without equilibrium points are considered, then these attractors are known as attractors without equilibria.

Definition 2.1. An attractor in a vector field with equilibria is called a self-excited attractor if its basin of attraction intersects with any open neighborhood of an unstable fixed point. Otherwise, it is called a hidden attractor. An attractor in a vector without equilibria is called a non-self-excited attractors.

In this chapter, we address the generation of self-excited attractors via piecewise linear (PWL) system.

2.1 Chaotic behavior in PWL systems

Let $T : X \to X$, with $X \subset \mathbb{R}^n$ and $n \in \mathbb{Z}^+$, be a piecewise linear dynamical system whose dynamics is given by a family of subsystems of the form

$$\dot{\mathbf{x}} = A_\tau \mathbf{x} + B_\tau, \qquad (2.1)$$

where $\mathbf{x} = (x_1, \ldots, x_n)^T \in \mathbb{R}^n$ is the state vector, $A_\tau = \{a_{ij}^\tau\} \in \mathbb{R}^{n \times n}$, and $B_\tau = (\beta_{\tau 1}, \ldots, \beta_{\tau n})^T \in \mathbb{R}^n$ are the linear operators and constant real vectors of the τth-subsystems, respectively. The index $\tau \in \mathcal{I} = \{1, \ldots, \eta\}$ is given by a rule that switches the activation of a subsystem in order to determine the dynamics of the PWL system. Let X be a subset of $\mathbb{R}^n$ and $\mathcal{P} = \{P_1, \ldots, P_\eta\}$ ($\eta > 1$) be a finite partition of X, that is, $X = \bigcup_{1 \leq i \leq \eta} P_i$, and $P_i \cap P_j = \emptyset$ for $i \neq j$. Each element of the set $\mathcal{P}$ is called an atom.

The selection of the index τ can be given according to a predefined itinerary and controlling by time; or by requiring that τ takes its value according to the state variable $\mathbf{x}$ depending upon which atom of a finite partition of the state-space $\mathcal{P} = \{P_1, \ldots, P_\eta\}$ ($\eta \in \mathbb{Z}^+$) a point is in.

An easy way to generate a partition $\mathcal{P}$ is to consider a vector $\mathbf{v} \in \mathbb{R}^n$ (with $\mathbf{v} \neq 0$) and a set of scalars $\delta_1 < \delta_2 < \cdots < \delta_{\eta-1}$ such that each $P_i = \{\mathbf{x} \in \mathbb{R}^n : \delta_{i-1} \leq \mathbf{v}^T \mathbf{x} < \delta_i\}$, with $i = 2, \ldots, \eta - 1$, $P_1 = \{\mathbf{x} \in \mathbb{R}^n : \mathbf{v}^T \mathbf{x} < \delta_1\}$, and $P_\eta = \{\mathbf{x} \in \mathbb{R}^n : \delta_{\eta-1} \leq \mathbf{v}^T \mathbf{x}\}$. We call the hyperplanes $\mathbf{v}^T \mathbf{x} = \delta_i$ ($i = 1, \ldots, \eta - 1$) the switching surfaces. Without loss of generality, we assume that the hyperplanes $\mathbf{v}^T \mathbf{x} = \delta_i$ (for $i = 1, 2, \ldots, \eta - 1$) are defined with $\mathbf{v} = (1, 0, \ldots, 0)^T \in \mathbb{R}^n$.

We consider a piecewise linear system $(T, \mathcal{P})$, such that its restriction to each atom P_i has a fixed point $\mathbf{x}_i^*$ i.e. $T(\mathbf{x}_i^*) = 0$ for one $\mathbf{x}_i^* \in P_i$ ($i \in \mathcal{I}$). Clearly $\mathbf{x}_i^* = -A_\tau^{-1} B_\tau$. We assume that the switching signal depends on the state variable and is defined as follows:

Definition 2.2. Let $\mathcal{I} = \{1, 2, \ldots, \eta\}$ be an index set that labels each element of the family of the subsystems (2.1). A function $\kappa : \mathbb{R}^n \to \mathcal{I} = \{1, \ldots, \eta\}$ of the form

$$\kappa(\mathbf{x}) = \begin{cases} 1, & \text{if } \mathbf{x} \in P_1; \\ 2, & \text{if } \mathbf{x} \in P_2; \\ \vdots & \vdots \\ \eta, & \text{if } \mathbf{x} \in P_\eta; \end{cases} \tag{2.2}$$

is called a switching signal. Furthermore, if $\kappa(\mathbf{x}) = \tau_i \in \mathcal{I}$ is the value of the switching signal during the time interval $t \in [t_i, t_{i+1})$, then $\mathcal{S}(\mathbf{x}_0) = \{\tau_0, \tau_1, \ldots, \tau_m, \ldots\}$ gives the itinerary generated by $\kappa(\mathbf{x}_0)$ at $\mathbf{x}_0$ and, $\mathcal{S}(i, \mathbf{x}_0)$ is the element $\tau_i \in \mathcal{S}(\mathbf{x}_0)$ that occurs at time t_i, this defines a set of switching times $\Delta_t = \{t_0, t_1, \ldots, t_m, \ldots\}$.

Note that τ changes only when the orbit $\phi(t, \mathbf{x}_0)$ goes from one atom P_i to another P_j, $i \neq j$.

Definition 2.3. A η-PWL system is composed of two sets: $\mathbf{A} = \{A_1, \ldots, A_\eta\}$ and $\mathbf{B} = \{B_1, B_2, \ldots, B_\eta\}$, with $A_\tau = \{\alpha_{ij}^\tau\} \in \mathbb{R}^{n \times n}$ ($\alpha_{ij}^\tau \in \mathbb{R}$) and $B_\tau = (\beta_{\tau 1}, \ldots, \beta_{\tau n})^T \in \mathbb{R}^n$; and a switching signal $\kappa : \mathbb{R}^n \to \mathcal{I} = \{1, 2, \ldots, \eta\}$ so that:

$$\dot{\mathbf{x}} = \begin{cases} A_1 \mathbf{x} + B_1, \; if \; \kappa(\mathbf{x}) = 1; \\ A_2 \mathbf{x} + B_2, \; if \; \kappa(\mathbf{x}) = 2; \\ \quad \vdots \qquad\qquad \vdots \\ A_\eta \mathbf{x} + B_\eta, \; if \; \kappa(\mathbf{x}) = \eta. \end{cases} \tag{2.3}$$

We can rewrite (2.3) in a more compact form as:

$$\dot{\mathbf{x}} = A_{\kappa(\mathbf{x})} \mathbf{x} + B_{\kappa(\mathbf{x})}. \tag{2.4}$$

Definition 2.4. Two η_1-PWL and η_2-PWL systems are called quasi-symmetrical if they are governed by the same linear operator $A = A_i$ for all i but $\eta_1 \neq \eta_2$.

Now we assume that the dimension of each η-PWL system is $n = 3$ and that the eigenspectra of linear operators $A_\tau \in \mathbb{R}^{3 \times 3}$ have the following features: (a) one eigenvalue is a real number; and (b) two eigenvalues are complex conjugate numbers with non-zero imaginary part. There is an approach to generate dynamical systems based on these linear dissipative systems in the case where the complex eigenvalues and the real eignenvalue have mixed sign (sometimes called an unstable dissipative system (UDS) [Ontañón-García *et al.* (2014)]). In this paper we use a particular type of unstable dissipative system (UDS) called *Type I*:

Consider A of the form:

$$A_\tau = \begin{pmatrix} 0 & 1 & 0 \\ 0 & 0 & 1 \\ \alpha & \beta & \gamma \end{pmatrix}. \tag{2.5}$$

Definition 2.5. A subsystem (A_τ, B_τ) of the system (2.4) in $\mathbb{R}^3$ is said to be an UDS of *Type I* if the eigenvalues of the linear operator A_τ denoted by λ_i satisfy: $\sum_{i=1}^3 \lambda_i < 0$; λ_1 is a negative real eigenvalue and; the other two λ_2 and λ_3 are complex conjugate eigenvalues with positive real part. The system is an UDS of *Type II* if $\sum_{i=1}^3 \lambda_i < 0$, and one λ_i is a positive real eigenvalue and; the other two λ_i are complex conjugate eigenvalues with negative real part.

Proposition 2.1. *Consider the family of affine lineal systems* (2.3) *with lineal operator* A_τ *given by* (2.39) *with* $\alpha, \beta, \gamma \in \mathbf{R}$. *Let* $\{a, b, c\}$ *be a set of non-zero real numbers called control parameters. If* $\alpha = c(a^2 + b)$, $\beta = a^2 + b + 2ac$ *and* $\gamma = c - 2a$ *with* $b, c > 0$ *and* $a < c/2$, *then the system* (2.3) *is based on UDS Type I; on the other hand, if* $b > 0$ *and* $a, c < 0$, *then the system is based on UDS Type II.*

Proof. The characteristic polynomial of the lineal operator (2.39) is:

$$
\begin{aligned}
p(\lambda) &= \lambda^3 + \gamma\lambda^2 + \beta\lambda + \alpha \\
&= \lambda^3 + (c - 2a)\lambda^2 + (a^2 + b + 2ac)\lambda + (ca^2 + cb) \\
&= (\lambda + c)(\lambda^2 - 2a\lambda + (a^2 + b)).
\end{aligned}
$$

The roots of $p(\lambda)$ give the following expressions for the eigenspectra $\Lambda = \{\lambda_1, \lambda_2, \lambda_3\}$ of (2.39): $\lambda_1 = -c$ and $\lambda_{2,3} = a \pm i\sqrt{b}$. Note that $\lambda_1 < 0$ and $\sum_{i=1}^{3} \lambda_i = -c + 2a < 0$ if $a < c/2$ and $c > 0$. Then, according to Definition 2.5 the system (2.3) is UDS *Type I*. On the other hand, if $a, c < 0$, then $\lambda_1 > 0$ and the above summation is still negative since $a > c/2$, which implies that the system is UDS *Type II*. $\qquad\square$

Proposition 2.2. *Consider the family of affine lineal systems given by* (2.3), *the lineal operator* A *based on the jerk system* (2.39) *with* $\alpha, \beta, \gamma \in \mathbf{R}$. *If* $\alpha > 0$, $0 < \beta < \alpha/\gamma$ *and* $\gamma > 0$, *then the system* (2.3) *is based on UDS Type I.*

Proof. Suppose $\alpha, \gamma > 0$. Since, by definition, $-\gamma = Trace(A) = \sum_{i=1}^{3} \lambda_i < 0$, system (2.3) is dissipative. Additionally, with $\alpha = det(A)$ the system (2.3) has a saddle equilibrium, which is determined by the characteristic polynomial of the lineal operator (2.39) is:

$$
p(\lambda) = \lambda^3 + \gamma\lambda^2 + \beta\lambda + \alpha,
$$

which for $\beta < \alpha/\gamma$, according with Hurwitz polynomial criterion, implies unstability. Due to α, β and γ are positive and according to Descartes' rule of signs the characteristic polynomial has no positive roots, so it has only one negative root due to the equilibrium point is a saddle. Then the eigenspectra is given by one negative real eigenvalue and a pair of complex conjugate eigenvalues with positive real part. $\qquad\square$

To each $\tau \in \mathcal{I}$ is associated an atom $P_\tau \subset \mathbb{R}^n$, containing an equilibrium point $\mathbf{x}_\tau^* = -A^{-1}B_\tau$ which has a one-dimensional stable manifold $E^s = Span\{\bar{v}_j \in \mathbb{R}^3 : \alpha_j < 0\}$ and a two-dimensional unstable manifold $E^u = Span\{\bar{v}_j \in \mathbb{R}^3 : \alpha_j > 0\}$, with $\bar{v}_j$ an eigenvector of the linear operator A and

$\lambda_j = \alpha_j + i\beta_j$ its corresponding eigenvalue; i.e., it is a saddle equilibrium point. We are interested in bounded flows which are generated by quasi-symmetrical η-PWL systems such that for any initial condition $\mathbf{x}_0 \in \mathbb{R}^3$, the orbit $\phi(t, \mathbf{x}_0)$ of the η-PWL system (2.4) limits a one-spiral trajectory in the atom P_τ called a scroll. The orbit escapes from one atom to other due to the unstable manifold in each atom. In this context, the system η-PWL (2.4) can display various multi-scroll attractors as a result of a combination of several unstable one-spiral trajectories, while the switching between regions is governed by the function (2.2).

Definition 2.6. The scroll-degree of a η-PWL system (2.4) based on UDS *Type I* is the maximum number of scrolls that the PWL system can display in the attractor.

In this section we analyze some of the systems that generate n-scrolls by different methods as hysteresis, step function and saturation. We focus on those that describe their experiments using the ODE written in the jerky form (2.6). This type of system, was implemented in [Elwakil *et al.* (2000); Yalçin *et al.* (2002); Xie *et al.* (2008); Lü *et al.* (2004); Lu *et al.* (2004); Deng and Lü (2007)], and the characteristic polynomial takes the form (2.7) regardless of the method used (hysteresis [Lü *et al.* (2004)], saturation [Lu *et al.* (2004)], step function [Yalçin *et al.* (2002)]).

The aforementioned chaotic approaches through PWL systems have used the particular case of the linear ordinary differential equation (ODE) written in the jerky form as $d^3x/dt^3 + a_{33}d^2x/dt^2 + a_{32}dx/dt + a_{31}x + \beta_3 = 0$, representing the state space equations of (2.4), where the matrix A and the vector B are described as follows:

$$A = \begin{pmatrix} 0 & 1 & 0 \\ 0 & 0 & 1 \\ -a_{31} & -a_{32} & -a_{33} \end{pmatrix} ; \quad B = \begin{pmatrix} 0 \\ 0 \\ \beta_3 \end{pmatrix}, \tag{2.6}$$

where the coefficients $a_{31}, a_{32}, a_{33}, \beta_3 \in \mathbb{R}$ may be any arbitrary scalars that satisfy the Definition 2.5. The characteristic polynomial of matrix A given by (2.6) takes the following form:

$$\lambda^3 + a_{33}\lambda^2 + a_{32}\lambda + a_{31}. \tag{2.7}$$

For instance, if the coefficient $a_{31} \in \mathbb{R}$ is varied and the others two coefficients are set at $a_{32} = 1$, $a_{33} = 1$ we can find two intervals that fulfill the definition 2.5. Therefore, for this case the coefficient a_{31} has to assure the system will be UDS-I or UDS-II. Figure 2.1 shows the location of the roots, and we can observe that the UDS's-II are given for $a_{31} < 0$, and the UDS's-I for $a_{31} > 1$. The system has a sink for $0 < a_{31} < 1$. In this way if we set $a_{31} = 1.5$ then we assure a UDS-I, with these values the eigenvalues result in $\lambda_1 = -1.20$, $\lambda_{2,3} = 0.10 \pm 1.11i$, which satisfy Definition 2.5 for UDS-I.

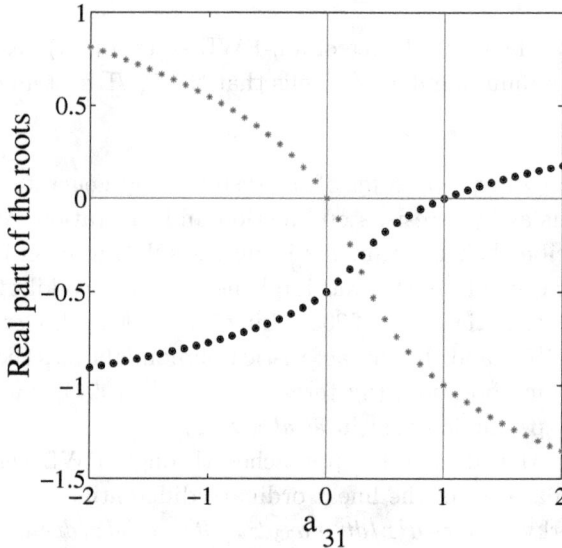

Fig. 2.1   The location of the roots: UDS-II for $a_{31} < 0$; UDS-I for $a_{31} > 1$.

Saturation:

A saturation function was implemented in [Lu *et al.* (2004)], this approach may generate 1D n-scroll, 2D $n \times m$-grid scroll, and 3D $n \times m \times l$-grid scroll chaotic attractors. The saturation function implemented in [Lu *et al.* (2004)] can be described with UDS-I and -II by means of Eq. (2.4), where $\mathcal{P} = \{P_1 = \{\mathbf{x}|x_1 < -1\}, P_2 = \{\mathbf{x}| -1 \leq x_1 \leq 1\}, P_3 = \{\mathbf{x}|x_1 > 1\}\}$, $\mathbf{B} = \{B_1 = (0,0,-7)^T, B_2 = (0,0,0)^T, B_3 = (0,0,7)^T\}$. and a switching signal taking the following form:

$$\kappa(\mathbf{x}) = \begin{cases} 1, \text{ if } \mathbf{x} \in P_1, \\ 2, \text{ if } \mathbf{x} \in P_2, \\ 3, \text{ if } \mathbf{x} \in P_3. \end{cases} \tag{2.8}$$

According to the specific values described in [Lu *et al.* (2004)], the matrices A_i, with $i = 1, 3$, in (2.6) takes the following values $a_{31} = a_{32} = a_{33} = 0.7$, for $\mathbf{x} \in P_1 \cup P_3$. The system presents a double scroll chaotic attractor with eigenvalues $\lambda_1 = -0.848$, $\lambda_{2,3} = 0.074 \pm 0.905i$, satisfying Definition 2.5 for UDS-I. The equilibrium points are located at $(\pm 10, 0, 0)^T$, this may be appreciated in Fig. 2.2. For $\mathbf{x} \in P_2$, the matrix A_2 in (2.6) takes the following values $\alpha_{31} = 6.3$, $\alpha_{32} = \alpha_{33} = 0.7$. Thus, A_2 has the following eigenvalues $\lambda_1 = 1.530$, $\lambda_{2,3} = -1.115 \pm 1.694i$, satisfying Definition 2.5 for UDS-II. The equilibrium point is located at the origin $(0, 0, 0)^T$. Here there are three equilibrium points and the flow of the system crosses from one to another because the system contains UDS-I and -II. In this way it is possible to describe all systems presented in [Lu *et al.* (2004)].

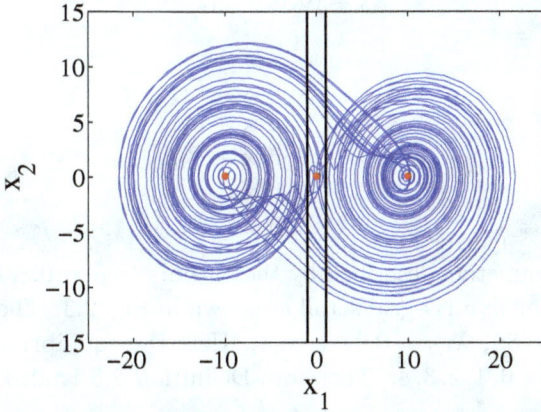

Fig. 2.2   The projection onto the plane (x_1, x_2) of the attractors generated by a saturation serie. The equilibrium points are marked with red dots, and the switching surface with black lines at $x_1 = \pm 1$.

The mechanism of generation of the attractor presented in Fig. 2.2 in terms of UDS is given as follows:

$$UDS = \begin{cases} \text{type I,} & \text{if } \mathbf{x} \in P_1, \\ \text{type II,} & \text{if } \mathbf{x} \in P_2, \\ \text{type I,} & \text{if } \mathbf{x} \in P_3. \end{cases} \tag{2.9}$$

Notice that the Chua's system can be described in a similar way that saturated functions for generation of multi-scroll attractors. Here there are some open questions, for example about the length of the middle region that contains the UDS-II in order to maintain the attractor, another question is about the minimum proximity of the equilibrium point to the commutation surface in order to generate a scroll around it.

Step function:

[Yalçin *et al.* (2002)] made an implementation of a step function from which they can generate 1D, 2D, 3D-grid scroll attractors. The step function depicted in Eqs. (2.6) and (2.7) in [Yalçin *et al.* (2002)], may be interpreted with UDS-I as follows:

$$\mathcal{P} = \{P_1 = \{\mathbf{x}|\ x_1 < 0.5\}, P_2 = \{\mathbf{x}|\ 0.5 \leq x_1 < 1.5\}, P_3 = \{\mathbf{x}|\ 1.5 \leq x_1 < 2.5\},$$
$$P_4 = \{\mathbf{x}|\ 2.5 \leq x_1 < 3.5\}, P_5 = \{\mathbf{x}|\ x_1 > 3.5\}\},$$

$$\mathbf{B} = \{B_1 = (0,0,0)^T,\ B_2 = (0,0,1)^T,\ B_3 = (0,0,2)^T,\ B_4 = (0,0,3)^T,$$
$$B_5 = (0,0,4)^T\},$$

$$\kappa(\mathbf{x}) = \begin{cases} 1, \text{ if } \mathbf{x} \in P_1, \\ 2, \text{ if } \mathbf{x} \in P_2, \\ 3, \text{ if } \mathbf{x} \in P_3, \\ 4, \text{ if } \mathbf{x} \in P_4, \\ 5, \text{ if } \mathbf{x} \in P_5. \end{cases} \tag{2.10}$$

Using the parameters described by the authors ($\alpha_{31} = 0.8$, $\alpha_{32} = \alpha_{33} = 1$) according to the five 1D-grid scroll as shown in Fig. 2.3. The eigenvalues result in $\lambda_1 = -0.89$, $\lambda_{1,2} = 0.04 \pm 0.94i$. Here the equilibrium points are $(i,0,0)^T$ with $i = 0,1,2,3,4$. Therefore Definition 2.5 is also satisfied for UDS-I.

Hysteresis:

[Lü *et al.* (2004)], implemented a hysteresis series to obtain multiscroll chaotic attractors in 1D *n*-scroll, 2D $n \times m$-grid scroll, and 3D $n \times m \times l$-grid scroll. The hysteresis series is given by Eqs. (2.4) and (2.6) in [Lü *et al.* (2004)], as follows:

$$h(x_1) = \begin{cases} 0, \text{ for } x_1 < 1, \\ 1, \text{ for } x_1 > 0, \end{cases} \tag{2.11}$$

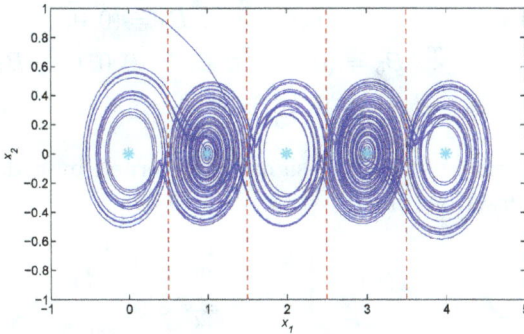

Fig. 2.3 The projection onto the plane (x_1, x_2) of the attractors generated by a step function. The equilibrium points are marked with asterisk, and the switching surface with a dotted line.

where $h(x_1)$ is the hysteresis function that is used to define the following hysteresis series,

$$h(x_1, p, q) = \sum_{i=1}^{p} h_{-i}(x_1) + \sum_{i=1}^{q} h_i(x_1), \qquad (2.12)$$

where p and q are positive integers, and $h_i(x_1) = h(x_1 - i + 1)$ and $h_{-i}(x_1) = -h_i(x_1)$. Equation (2.12) can be recast as follows:

$$h(x_1, p, q) = \begin{cases} -p, & \text{for } x_1 < -p + 1, \\ i, & \text{for } \begin{cases} i - 1 < x_1 < i + 1, \\ i = -p + 1, \ldots, q - 1, \end{cases} \\ q, & \text{for } x_1 > q - 1. \end{cases} \qquad (2.13)$$

Using Eq. (2.2), it is possible to generate multi-scroll attractors based on hysteresis series. This approach may be explained with the UDS definition. Considering the parameters described by the authors, matrix A in (2.6) takes the following values: $\alpha_{31} = 0.8$, $\alpha_{32} = 0.72$, $\alpha_{33} = 0.6$. Therefore the eigenvalues result in $\lambda_1 = -0.85$, $\lambda_{2,3} = 0.12 \pm 0.95i$, where Definition 2.5 is satisfied for UDS-I. The equilibrium points for the 1D n-scroll are located along the x-axis. Now the partition of the space is dynamic, i.e., it is given by two sets

$$\mathcal{P} = \{P_1 = \{\mathbf{x}|\, x_1 < -3\},\ P_2 = \{\mathbf{x}|\, -3 < x_1 < -2\},$$
$$P_3 = \{\mathbf{x}|\, -2 < x_1 < -1\},\ P_4 = \{\mathbf{x}|\, -1 < x_1 < 0\},$$
$$P_5 = \{\mathbf{x}|\, 0 < x_1 < 1\},\ P_6 = \{\mathbf{x}|\, 1 < x_1 < 2\},$$
$$P_7 = \{\mathbf{x}|\, 2 < x_1 < 3\},\ P_8 = \{\mathbf{x}|\, x_1 > 3\}\},$$

$$\mathbf{B} = \{B_1 = (0,0,-4)^T, \ B_2 = (0,0,-3)^T, \ B_3 = (0,0,-2)^T,$$
$$B_4 = (0,0,-1)^T, \ B_5 = (0,0,0)^T, \ B_6 = (0,0,1)^T, \ B_7 = (0,0,2)^T,$$
$$B_8 = (0,0,3)^T, B_9 = (0,0,4)^T\}.$$

The corresponding switching signal is governed by a derivative which takes the form described next:

$$\kappa(\mathbf{x}) = \begin{cases} \kappa_+, \ \text{if } \frac{dx_1}{dt} > 0, \\ \kappa_-, \ \text{if } \frac{dx_1}{dt} < 0, \end{cases} \qquad (2.14)$$

where

$$\kappa_+ = \begin{cases} 1, \ \text{if } \mathbf{x} \in P_1; \\ 2, \ \text{if } \mathbf{x} \in P_2, \\ 3, \ \text{if } \mathbf{x} \in P_3, \\ 4, \ \text{if } \mathbf{x} \in P_4, \\ 5, \ \text{if } \mathbf{x} \in P_5, \\ 6, \ \text{if } \mathbf{x} \in P_6, \\ 7, \ \text{if } \mathbf{x} \in P_7, \\ 8, \ \text{if } \mathbf{x} \in P_8, \end{cases} \qquad (2.15)$$

and

$$\kappa_- = \begin{cases} 2, \ \text{if } \mathbf{x} \in P_2; \\ 3, \ \text{if } \mathbf{x} \in P_3, \\ 4, \ \text{if } \mathbf{x} \in P_4, \\ 5, \ \text{if } \mathbf{x} \in P_5, \\ 6, \ \text{if } \mathbf{x} \in P_6, \\ 7, \ \text{if } \mathbf{x} \in P_7, \\ 8, \ \text{if } \mathbf{x} \in P_8, \\ 9, \ \text{if } \mathbf{x} \in P_9. \end{cases} \qquad (2.16)$$

From this it may be concluded that the hysteresis series acts similarly to the switching signal described previously in (2.2). Equilibrium points are being introduced and the system is forced to oscillate around them. Since both systems are UDS-I, the orbit escapes by means of the unstable manifold E^u until it reaches the commutation surface and generates a change to other equilibrium point. This may be seen in Fig. 2.4.

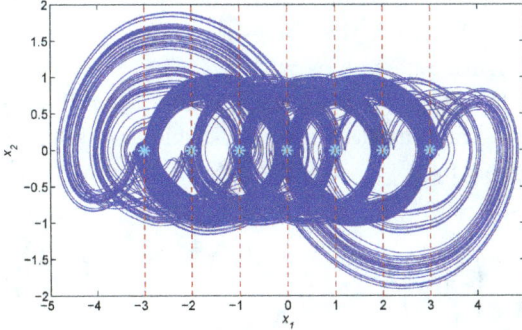

Fig. 2.4 The projection of the attractor onto the (x_1, x_2) plane generated by a hysteresis serie. The equilibrium points are marked with asterisk, and the switching surface with a dotted line.

2.2 Heteroclinic chaos via PWL systems

To introduce the approach, let us first consider a partition $\mathcal{P}$ of the metric space $X \subset \mathbb{R}^3$, endowed with the Euclidean metric d. Let $\mathcal{P} = \{P_1, \dots, P_\eta\}$ ($\eta > 1$) be a finite partition of X. Each atom contains a saddle equilibrium point and there is an orbit that connect two saddle equilibria of adjacent atoms. Remember that a heteroclinic orbit is a path that joins two equilibrium points in the phase space. Similarly, a homoclinic orbit is a path that starts and ends at the same equilibrium point.

Let $T : X \to X$, with $X \subset \mathbb{R}^3$, be a piecewise-linear dynamical system whose dynamics is given by a family of subsystems of the form

$$\dot{\mathbf{x}} = A\mathbf{x} + f(\mathbf{x})B, \qquad (2.17)$$

where $\mathbf{x} = (x_1, x_2, x_3)^T \in \mathbb{R}^3$ is the state vector, and $A = \{\alpha_{i,j}\} \in \mathbb{R}^{3\times3}$ is a linear operator that is defined as follows:

$$A = \begin{pmatrix} \dfrac{a}{3} + \dfrac{2c}{3} & b & \dfrac{2c}{3} - \dfrac{2a}{3} \\ -\dfrac{b}{3} & a & \dfrac{2b}{3} \\ \dfrac{c}{3} - \dfrac{a}{3} & -b & \dfrac{2a}{3} + \dfrac{c}{3} \end{pmatrix}, \qquad (2.18)$$

$B = (\beta_1, \beta_2, \beta_3)^T$ is a constant vector, and f is a functional. The vector $f(\mathbf{x})B$ is a constant vector in each atom P_i such that the equilibria are given by $\mathbf{x}^*_{eq_i} = (x^*_{1_{eq_i}}, x^*_{2_{eq_i}}, x^*_{3_{eq_i}})^T = -f(\mathbf{x})A^{-1}B \in P_i$, with $i = 1, \dots, \eta$. Oscillations of the flow around the equilibria $x^*_{eq_i}$ are desired.

Let us assign a negative real eigenvalue $\lambda_1 = c$ to the complexification of the operator $A(A_\mathbb{C})$ with the corresponding eigenvector v_1, and a pair of complex conjugate eigenvalues with positive real part $\lambda_2 = a + ib$ and $\lambda_3 = a - ib$ with the corresponding eigenvectors v_2 and v_3. Additionally, we restrict $b/a \geq 10$. Thus the stable and unstable manifolds of an equilibrium point x_{eq}^* are given by $E_{\mathbf{x}_{eqi}^*}^s = \{\mathbf{x} + \mathbf{x}_{eqi}^* : \mathbf{x} \in \mathrm{span}\{v_1\}\}$ and $E_{\mathbf{x}_{eqi}^*}^u = \{\mathbf{x} + \mathbf{x}_{eqi}^* : \mathbf{x} \in \mathrm{span}\{v_2, v_3\}\}$, where v_1, v_2 and v_3 are given as follows:

$$v_1 = \begin{pmatrix} 1 \\ 0 \\ 1/2 \end{pmatrix}, \quad v_2 = \begin{pmatrix} 0 \\ -1 \\ 0 \end{pmatrix}, \quad v_3 = \begin{pmatrix} -1 \\ 0 \\ 1 \end{pmatrix}. \tag{2.19}$$

We also denoted the closure of a set P_i as $cl(P_i)$. Thus, for each pair of atoms P_i and P_j, $i \neq j$, if $cl(P_i) \cap cl(P_j) \neq \emptyset$ then these atoms are adjacent and the switching surface between them is given by the intersection, i.e., $SW_{ij} = cl(P_i) \cap cl(P_j)$.

Each SW_{ij} has associated an equation of the form $\hat{A}x_1 + \hat{B}x_2 + \hat{C}x_3 + D = \mathbf{N_{12}} \cdot \mathbf{x}^\mathbf{T} + D = 0$, where $\hat{A} > 0$ and $\mathbf{N_{12}} = (\hat{A}, \hat{B}, \hat{C})$ is the normal vector. To generate a heteroclinic orbit, at least two equilibria are required, therefore, consider a partition with two atoms $\mathcal{P} = \{P_1, P_2\}$, that are defined as follows.

$$\begin{aligned} P_1 &= \{\mathbf{x} \in \mathbb{R}^3 : x_3 > 0, \ \mathbf{N_{12}} \cdot \mathbf{x}^\mathbf{T} \leq -D\} \\ &\cup \{\mathbf{x} \in \mathbb{R}^3 : x_3 \leq 0, \ \mathbf{N_{12}} \cdot \mathbf{x}^\mathbf{T} < -D\}, \\ P_2 &= \{\mathbf{x} \in \mathbb{R}^3 : x_3 > 0, \ \mathbf{N_{12}} \cdot \mathbf{x}^\mathbf{T} > -D\} \\ &\cup \{\mathbf{x} \in \mathbb{R}^3 : x_3 \leq 0, \ \mathbf{N_{12}} \cdot \mathbf{x}^\mathbf{T} \geq -D\}. \end{aligned} \tag{2.20}$$

Remark 2.1. The divergence of the PWL system (2.17) considering the linear operator A given by (2.18) is $\nabla = 2a + c$, so the system is dissipative in each atom of the partition $\mathcal{P}$ if $2a < |c|$.

With the atoms of a $\mathcal{P}$ partition containing a saddle equilibrium point in each of them as defined above, it is possible to generate heteroclinic orbits. The constant vector $B \in \mathbb{R}^3$ is defined as follows:

$$B = \begin{pmatrix} -\dfrac{a}{3} - \dfrac{2c}{3} \\ \dfrac{b}{3} \\ \dfrac{a}{3} - \dfrac{c}{3} \end{pmatrix}, \tag{2.21}$$

and the functional f is given by

$$f(\mathbf{x}) = \begin{cases} -\alpha, & \mathbf{x} \in P_1; \\ \alpha, & \mathbf{x} \in P_2; \end{cases} \tag{2.22}$$

with $\alpha > 0$. So the equilibria are at $\mathbf{x}^*_{eq_1} = (-\alpha, 0, 0)^T \in P_1$ and $\mathbf{x}^*_{eq_2} = (\alpha, 0, 0)^T \in P_2$, and the stable and the unstable manifolds are given by

$$W^s_{\mathbf{x}^*_{eq_1}} = \{\mathbf{x} \in P_1 \subset \mathbb{R}^3 | x_1 + \alpha = 2x_3, x_2 = 0\},$$

$$W^u_{\mathbf{x}^*_{eq_1}} = \{\mathbf{x} \in P_1 \subset \mathbb{R}^3 | x_1 + x_3 = -\alpha\},$$

$$W^s_{\mathbf{x}^*_{eq_2}} = \{\mathbf{x} \in P_2 \subset \mathbb{R}^3 | x_1 - \alpha = 2x_3, x_2 = 0\},$$

$$W^u_{\mathbf{x}^*_{eq_2}} = \{\mathbf{x} \in P_2 \subset \mathbb{R}^3 | x_1 + x_3 = \alpha\}.$$

Proposition 2.3 (**[Escalante-González and Campos (2020c); Escalante-González and Campos-Cantón (2019)]**). *The hyperbolic system given by* (2.17), (2.18), (2.21) *and* (2.22) *generates a pair of heteroclinic orbits if the switching surface between the atoms P_1 and P_2 is given by the plane $SW_{12} = \{\mathbf{x} \in \mathbb{R}^3 : 2x_1 - x_3 = 0\}$.*

Proof. The points where the stable and unstable manifolds intersect at SW are given by

$$\mathbf{x}_{in_1} = cl(W^s_{\mathbf{x}^*_{eq_1}}) \cap cl(W^u_{\mathbf{x}^*_{eq_2}}) = \left(\frac{\alpha}{3}, 0, \frac{2\alpha}{3}\right)^T.$$

$$\mathbf{x}_{in_2} = cl(W^s_{\mathbf{x}^*_{eq_2}}) \cap cl(W^u_{\mathbf{x}^*_{eq_1}}) = \left(-\frac{\alpha}{3}, 0, -\frac{2\alpha}{3}\right)^T.$$

These points $\mathbf{x}_{in_1}$ and $\mathbf{x}_{in_2}$ belong to SW_{12} and $\mathbf{x}_{in_1} \in P_1$ and $\mathbf{x}_{in_2} \in P_2$. Because these points $\mathbf{x}_{in_1}$ and $\mathbf{x}_{in_2}$ belong to the stable manifolds $W^s_{\mathbf{x}^*_{eq_1}}$ and $W^s_{\mathbf{x}^*_{eq_2}}$, respectively, they are points whose trajectories remain in atoms P_1 and P_2, respectively. Thus the heteroclinic orbits are defined as

$$HO_1 = \{\mathbf{x} \in \varphi(\mathbf{x}_{in_1}, t) : t \in (-\infty, \infty)\},$$

$$HO_2 = \{\mathbf{x} \in \varphi(\mathbf{x}_{in_2}, t) : t \in (-\infty, \infty)\}. \qquad \square$$

For the system given by (2.17), (2.18), (2.21) and (2.22), it is possible to find several points $\mathbf{x}_0 \in HO_i$ such that $|\mathbf{x}_{eq_i} - \mathbf{x}_0| < \epsilon$ with ϵ arbitrarily small and $i = 1, 2$. Thus, one can find initial conditions for the simulation of the heteroclinic orbits as close to the equilibria as desired. One example of initial condition formula for P_1 is

$$\mathbf{x}_0^1 = \begin{pmatrix} \frac{2}{3}\alpha e^{-\frac{2k\alpha\pi}{b}} - \alpha \\ 0 \\ -\frac{2}{3}\alpha e^{-\frac{2k\alpha\pi}{b}} \end{pmatrix}, \qquad (2.23)$$

and for P_2

$$\mathbf{x}_0^2 = \begin{pmatrix} -\frac{2}{3}\alpha e^{-\frac{2k a \pi}{b}} + \alpha \\ 0 \\ \frac{2}{3}\alpha e^{-\frac{2k a \pi}{b}} \end{pmatrix}, \qquad (2.24)$$

with $k \in \mathbb{Z}^+$.

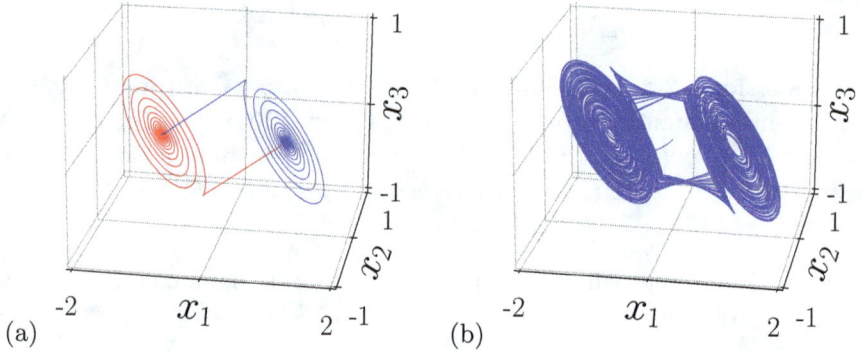

Fig. 2.5   In (a) the heteroclinic loop of the system (2.17), (2.18), (2.21) and (2.22) with the switching surface $\{x \in \mathbb{R}^3 : 2x_1 - x_3 = 0\}$, the parameters $a = 0.2, b = 5, c = -3, \alpha = 1$, and the initial conditions $\mathbf{x}_{01} = (-0.9999976751050959, 0, -2.3248949041393315e - 6)^T$ (red) and $\mathbf{x}_{02} = (0.9999976751050959, 0, 2.3248949041393315e - 6)^T$ (blue), and in (b) a double-scroll attractor that emerges from an heteroclinic orbit using the following initial condition $\mathbf{x}_0 = (0, 0, 0)^T$ and the same parameters.

Example 2.1. Consider the system (2.17), (2.18), (2.21) and (2.22) with $SW_{12} = \{\mathbf{x} \in \mathbb{R}^3 : 2x_1 - x_3 = 0\}$ and the parameters $a = 0.2, b = 5, c = -3, \alpha = 1$.

The above defined system fulfills the proposition 2.3, so it presents a heteroclinic orbit. From (2.23) and (2.24) two initial conditions $\mathbf{x}_{01} = (-0.9999976751050959, 0, -2.3248949041393315e - 6)^T$ and $\mathbf{x}_{02} = (0.9999976751050959, 0, 2.3248949041393315e - 6)^T$ are chosen with $k = 50$ to simulate the two heteroclinic orbits shown in the Fig. 2.5(a). A double-scroll attractor with heteroclinic chaos is generated and it is shown in the Fig. 2.5(a) for the initial condition $\mathbf{x}_0 = (0, 0, 0)^T$.

The unstable manifolds $W^u_{\mathbf{x}^*_{eq1}} = \{\mathbf{x} \in \mathbb{R}^3 : x_1 + x_3 + 1 = 0\}$ and $W^u_{\mathbf{x}^* eq2} = \{\mathbf{x} \in \mathbb{R}^3 : x_1 + x_3 - 1 = 0\}$ and the stable manifolds $W^s_{\mathbf{x}^*_{eq1}} = \{\mathbf{x} \in \mathbb{R}^3 : \frac{x_1+1}{2} = x_3; x_2 = 0\}$ and $W^s_{\mathbf{x}^*_{eq2}} = \{\mathbf{x} \in \mathbb{R}^3 : \frac{x_1-1}{2} = x_3; x_2 = 0\}$. The intersection points are given by $cl(W^s_{\mathbf{x}^*_{eq2}}) \cap cl(W^u_{\mathbf{x}^*_{eq1}}) = (-\frac{1}{3}, 0, -\frac{2}{3})^T$, $cl(W^s_{\mathbf{x}^*_{eq1}}) \cap cl(W^u_{\mathbf{x}^*_{eq2}}) = (\frac{1}{3}, 0, \frac{2}{3})^T$.

2.3   One direction grid scrolls attractor and its design

It is possible to generate multi-scroll attractors based on multiple hetero-clinic orbits. So in this Section 2.3 we consider more than two hyperbolic sets in the partition with the aim of studying the existence of heterocyclic cycles and one direction grid scrolls attractor.

Proposition 2.4 ([**Escalante-González and Campos-Cantón (2019)**]). *If the partition P contains more than two atoms $\{P_1, P_2, \ldots, P_k\}$, with $2 < k \in \mathbb{Z}^+$, and each atom is a hyperbolic set defined as above. Furthermore, the atoms by pairs P_i and P_{i+1} fulfill the Proposition 2.3. Then the system generates $2(k-1)$ heteroclinic orbits.*

Proof. A direct consequence of the Proposition 2.3. □

We start by considering a partition with four atoms $\mathcal{P} = \{P_1, P_2, P_3, P_4\}$ along with the piecewise linear dynamical system (2.17), with A and B given by (2.18) and (2.21), respectively, thus the functional $f(\mathbf{x})$ is defined in the four atoms as follows:

$$f(\mathbf{x}) = \begin{cases} -3\alpha, & \mathbf{x} \in P_1; \\ -\alpha, & \mathbf{x} \in P_2; \\ \alpha, & \mathbf{x} \in P_3; \\ 3\alpha, & \mathbf{x} \in P_4; \end{cases} \tag{2.25}$$

where $\alpha \in \mathbb{R}$. The equilibria are at:

$$\mathbf{x}^*_{eq_1} = \begin{bmatrix} -3\alpha \\ 0 \\ 0 \end{bmatrix}, \quad \mathbf{x}^*_{eq_2} = \begin{bmatrix} -\alpha \\ 0 \\ 0 \end{bmatrix}, \quad \mathbf{x}^*_{eq_3} = \begin{bmatrix} \alpha \\ 0 \\ 0 \end{bmatrix}, \quad \mathbf{x}^*_{eq_4} = \begin{bmatrix} 3\alpha \\ 0 \\ 0 \end{bmatrix}, \tag{2.26}$$

so $\mathbf{x}^*_{eq_1} \in P_1$, $\mathbf{x}^*_{eq_2} \in P_2$, $\mathbf{x}^*_{eq_3} \in P_3$ and $\mathbf{x}^*_{eq_4} \in P_4$. The location of the equilibria according to the parameter α are on the x_1 axis.

The switching surfaces are given by:

$$\begin{aligned} SW_{12} &= cl(P_1) \cap cl(P_2) = \{\mathbf{x} \in \mathbb{R}^3 : 2x_1 - x_3 = -4\alpha\}, \\ SW_{23} &= cl(P_2) \cap cl(P_3) = \{\mathbf{x} \in \mathbb{R}^3 : 2x_1 - x_3 = 0\}, \\ SW_{34} &= cl(P_3) \cap cl(P_4) = \{\mathbf{x} \in \mathbb{R}^3 : 2x_1 - x_3 = 4\alpha\}, \end{aligned} \tag{2.27}$$

which fulfill that

$$\begin{aligned} SW_{i(i+1)} \cap \{\mathbf{x} \in \mathbb{R}^3 : x_3 > 0\} &\in P_i, \\ SW_{i(i+1)} \cap \{\mathbf{x} \in \mathbb{R}^3 : x_3 \le 0\} &\in P_{i+1}. \end{aligned} \tag{2.28}$$

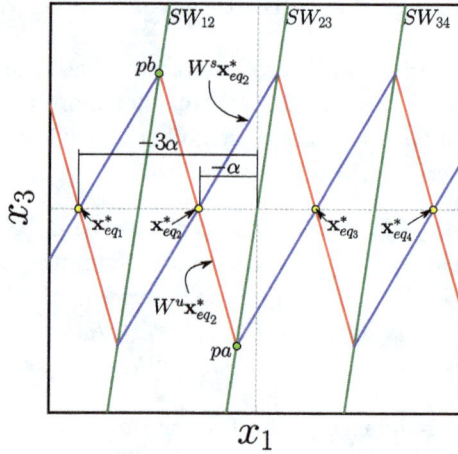

Fig. 2.6   Projection of the stable and unstable manifolds and switching planes onto the $x_1 - x_3$ plane. The diagram shows the location of the unstable manifold marked with red lines, the stable manifold marked with blue lines and switching planes marked with green lines.

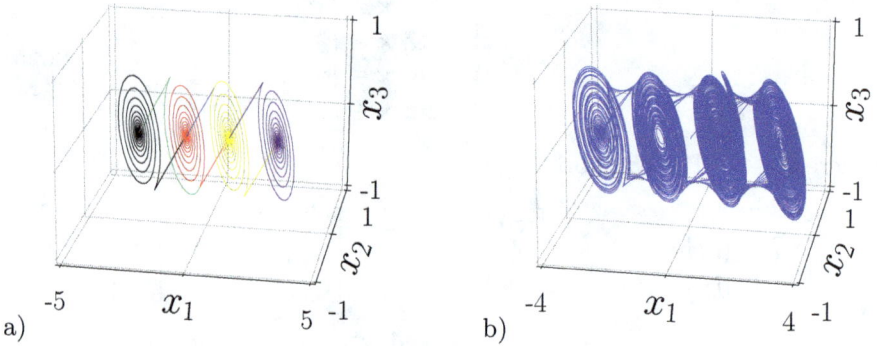

Fig. 2.7   Heteroclinic orbits of the system given by (2.17), (2.18), (2.21), (2.25) and (2.27) for the parameters $a = 0.2$, $b = 5$, $c = -3$, $\alpha = 1$ and $\gamma = 1$. (a) There are six heteroclinic orbits. (b) One directional grid chaotic attractor for $\mathbf{x}_0 = 0$.

This way of defining the switching surfaces provokes that the intersections between them and the stable manifolds contain a point, and the intersections between them and the unstable manifolds are the empty set, i.e., $W^u_{\mathbf{x}^*_{eq_1}} \cap SW_{12} = \emptyset$ and $W^s_{\mathbf{x}^*_{eq_1}} \cap SW_{12} \neq \emptyset$.

Let us define two points, $pb = W^s_{\mathbf{x}^*_{eq1}} \cap SW_{12}$ and $pa = W^s_{\mathbf{x}^*_{eq3}} \cap SW_{23}$ as shown in the Fig. 2.13. Then pa and pb are given as follows

$$pa = \begin{pmatrix} -\frac{(\alpha)}{3} \\ 0 \\ -\frac{2(\alpha)}{3} \end{pmatrix}, \ pb = \begin{pmatrix} -\frac{5\alpha}{3} \\ 0 \\ \frac{2\alpha}{3} \end{pmatrix}. \tag{2.29}$$

The set $cl(W^u_{\mathbf{x}^*_{eq2}}) \cap SW_{12}$ can be written as:

$$\{\mathbf{x} \in \mathbb{R}^3 : \ \mathbf{x} = (0, \epsilon, 0)^T + pb, \epsilon \in \mathbb{R}\}, \tag{2.30}$$

and the set $cl(W^u_{\mathbf{x}^*_{eq2}}) \cap SW_{23}$ can be written as:

$$\{\mathbf{x} \in \mathbb{R}^3 : \ \mathbf{x} = (0, \epsilon, 0)^T + pa, \epsilon \in \mathbb{R}\}. \tag{2.31}$$

The points pa and pb warranty the existence of heteroclinic orbits. Therefore, the hyperbolic system given by (2.17), (2.18), (2.21) and (2.25) with the switching surfaces given in (2.27) generates six heteroclinic orbits.

Figure 2.7(a) shows six heteroclinic orbits of the system given by (2.17), (2.18), (2.21), (2.25) and (2.27) for the parameters $a = 0.2$, $b = 5$, $c = -3$, $\alpha = 1$ and $\gamma = 1$. And Fig. 2.7(b) shows a one directional grid chaotic attractor for $\mathbf{x}_0 = 0$.

2.3.1 *Function for the commutation surface displacement*

According with Definition 2.5 the following entries of the matrix A will be considered:

$$A = \begin{pmatrix} 0 & 1 & 0 \\ 0 & 0 & 1 \\ -10.5 & -7.0 & -0.7 \end{pmatrix}. \tag{2.32}$$

And for the affine vector B defined in Eq. (2.6), the value of b_3 will commute according to the value of x_1 as follows:

$$b_3(x_1) = \begin{cases} c_1, \text{ if } x_1 \geq x_{\pm i_{cs}}; \\ c_0, \text{ otherwise.} \end{cases} \tag{2.33}$$

Here $c_{0,1} \in \mathbb{R}$ determine the values of the equilibrium points since $\mathbf{X}^*_i = (b_3/a_{31}, 0, 0)^T = (c_i/10.5, 0, 0)^T | i = 0, 1$, and $x_{\pm i_{cs}}$ stands for the location of the commutation surface (defined from now and on as P_{cs}) given along the x_1 axis regarding the position that it takes with respect to the positive or negative axis and with $i \in \mathbb{Z}$. Considering

$c_0 = 0$ and $c_1 = 6.3$ the equilibria are located at $\mathbf{X}_0^* = (0, 0, 0)^T$ and $\mathbf{X}_1^* = (0.6, 0, 0)^T$ displacing only along the positive x_1 axis. In order to consider an equally distributed scrolling around the equilibria, the distance between the equilibrium points is calculated with the euclidean distance $\alpha(\mathbf{X}_0^*, \mathbf{X}_1^*) = \sqrt{(x_{10}^* - x_{11}^*)^2 + (x_{20}^* - x_{21}^*)^2 + (x_{30}^* - x_{31}^*)^2}$, resulting in $\alpha = 0.6$. Therefore, the commutation surface that generates two equally distributed scroll trajectories is given by $x_{1_{cs}} = \alpha(\mathbf{X}_0^*, \mathbf{X}_1^*)/2$ resulting in a commutation surface located at the $x_1 = 0.3$ plane. The eigenvalues and their corresponding eigenvectors result in:

$$\Lambda = \{\lambda_{1,2,3}\} = \{-1.3372, 0.3186 \pm i1.754\}.$$

$$\vartheta = \{\vartheta_{1,2,3}\} = \left\{ \begin{pmatrix} 0.4087 \\ -0.5466 \\ 0.7309 \end{pmatrix} \begin{pmatrix} -0.1160 \pm i0.0269 \\ 0.0379 \pm i0.3316 \\ 0.9351 \end{pmatrix} \right\}. \tag{2.34}$$

The commutation surface must be located taking into consideration the manifolds E^s, E^u and between $\mathbf{X}_{0,1}^*$, otherwise the trajectory of the system can escape and the scrolls are not formed (this is described in Fig. 2 and Fig. 3 of the references [Campos-Cantón *et al.* (2010); Ontañón-García *et al.* (2014)], respectively). The trajectory of the system given by the initial condition $\mathbf{X}_0 = (0.7, 0, 0)^T$ and Eqs. (2.1) with (2.32) and (2.33) presents a double scroll attractor Fig. 2.8. It is important to mention that the unstable manifold determined by the two complex conjugate eigenvalues is represented only by the real part of ϑ_2, however this manifold E^u corresponds to a plane, also these manifolds E^s and E^u end at the commutation surface but they were projected to the other domains in order to clarify their location with respect to the other manifolds.

Between both equilibria, the commutation surface at the plane $P_{cs} = \{(x_1, x_2, x_3)^T \in \mathbb{R}^3 | x_1 = 0.3\}$, divide the space in two domains $\mathcal{D}_{0,1}$ given by $\mathcal{D}_0 = \{(x_1, x_2, x_3)^T \in \mathbb{R}^3 | x_1 < 3\}$ and $\mathcal{D}_1 = \{(x_1, x_2, x_3)^T \in \mathbb{R}^3 | x_1 \geq 3\}$. Notice two important facts about the system, first that the scrolls are increasing their size due to the unstable manifold. Second, that the trajectory of the system oscillating around the equilibrium point $\mathbf{X}_0^*$ in $\mathfrak{A} \cap \mathcal{D}_0$ escapes from the domain $\mathcal{D}_0$ near the unstable manifold $E^u \subset \mathcal{D}_0$ where it crosses the commutation surface and it is attracted by the stable manifold $E^s \subset \mathcal{D}_1$ towards the equilibrium point $\mathbf{X}_1^*$ in the domain $\mathcal{D}_1$ located at the right side of the commutation surface. The process is repeated in the inverse way forming scrolls around each equilibrium point.

This property of the UDS can be easily extended for the generation of any number of scrolls along any of the axes if the above considerations are made when designing the system and locating equilibria along the axes in the following way [Ontanón-García and Campos-Cantón (2015); Jiménez-López *et al.* (2013)]. First, start considering $c_0 = 0$ (in case that the first equilibrium point is located at the origin) and the value of $c_1 \neq 0$. Thus the commutation surface P_{cs} that can result in two symmetric equilibrium points $\mathbf{X}_i^*, i = 0, 1$, is given as $P_{cs} = \{(x_1, x_2, x_3)^T \in \mathbb{R}^3 | x_1 = \alpha(\mathbf{X}_0^*, \mathbf{X}_1^*)/2\}$, resulting in the two symmetrical scrolls. Now in order to extend it along the x_1 axis, the distance $\alpha(\mathbf{X}_0^*, \mathbf{X}_1^*) = 0.6$ between the equilibria must be considered between adjacent equilibrium points in the system.

This idea of introducing more equilibria to the system can be done by considering the values $c_{\pm i} = \pm(a_{31}\alpha(\mathbf{X}_0^*, \mathbf{X}_1^*)k) = \pm(10.5\alpha(\mathbf{X}_0^*, \mathbf{X}_1^*)k)$, resulting in the consecutive equilibria along the x_1 axis $\mathbf{X}_{\pm i}^* = (\pm\alpha(\mathbf{X}_0^*, \mathbf{X}_1^*)k, 0, 0)^T$ with $i, k = 0, \ldots, n$ $(i, k \in \mathbb{Z}^+)$. The number of scrolls that are introduced in the 1D-grid along x_1 is $2n + 1$. The commutation surfaces also should be located according to the value of the distance, in this case they are at $x_{\pm i_{cs}} = \pm\alpha(\mathbf{X}_0^*, \mathbf{X}_1^*)(1+2k)/2$. Therefore the commutation law for a 5 scroll attractor is given by

$$b_3(x_1) = \begin{cases} c_2, & \text{if } x_1 \geq x_{2_{cs}}; \\ c_1, & \text{if } x_{1_{cs}} \leq x_1 < x_{2_{cs}}; \\ c_0, & \text{if } x_{-1_{cs}} < x_1 < x_{1_{cs}}; \\ c_{-1}, & \text{if } x_{-2_{cs}} < x_1 \leq x_{-1_{cs}}; \\ c_{-2}, & \text{if } x_1 \leq x_{-2_{cs}}. \end{cases} \qquad (2.35)$$

The resulting attractor can be appreciated in Figs. 2.8(c) and (d). Both figures present the commutation surfaces $P_{i_{cs}} = \{(x_1, x_2, x_3)^T \in \mathbb{R}^3 | x_1 = x_{i_{cs}}\}$ marked with black line, along with the equilibria $\mathbf{X}_{\pm i}^*$. Each equilibrium point also depicts the E^s and E^u manifolds regarding their eigenvectors which are parallel among them similar as in the two scroll attractor.

Adding more equilibria to the system can be easily implemented by using a step function instead of generating commutation surfaces manually, i.e., using an automatically step generating function as it has been applied in [Huerta-Cuellar *et al.* (2014a); Gilardi-Velázquez and Campos-Cantón (2018a)]. Here, a *round*(x) function will be implemented to simplify and automate this process. The function will be defined as follows:

$$round(x) = \begin{cases} \lceil x - 1/2 \rceil, & \text{for } x < 0; \\ \lfloor x + 1/2 \rfloor, & \text{for } x \geq 0. \end{cases} \qquad (2.36)$$

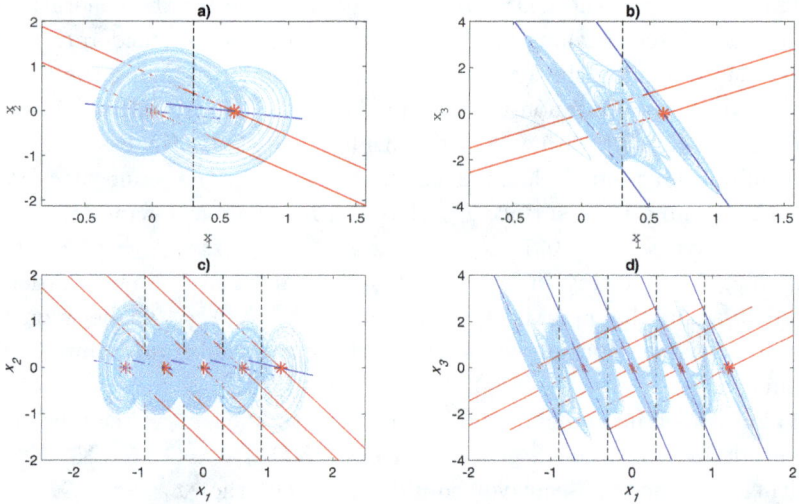

Fig. 2.8   Projections of the attractor given by Eqs. (2.1) with (2.32) and (2.33) onto the (x_1, x_2) plane for (a) and (c), and the (x_1, x_3) plane for (b) and (d). The commutation law given by Eq. (2.33) for (a) and (b), and Eq. (2.35) for (c) and (d). Marked with red asterisks the equilibria of the system and with black dashed line the commutation surface. The direction of the eigenvectors are marked with red and blue lines for the stable and unstable manifolds E^s and E^u, respectively.

For example, in order to have similar commutation surfaces and locations of equilibrium points that the ones described in Eq. (2.35), consider the commutation of the vector $\mathbf{B} = (0, 0, b_3)^T$ given by the following function:

$$b_3(x_1) = c * round(x_1/\alpha), \qquad (2.37)$$

where $c \in \mathbb{R}$ corresponds to the amplitude of the function which is similar to the variable c_i, and α corresponds to the length of the step given by the round function centered in the origin, this is depicted in the graph of Fig. 2.9. Notice that the value of α has the same representation as $\alpha(\mathbf{X}_0^*, \mathbf{X}_1^*)$. If the function (2.37) is considered as the commutation law of (2.6), then the commutation surfaces are located at every change of the steps, i.e., $x_{\pm i_{cs}} = \pm\alpha(1 + 2k)/2$ with $i, k = 1, \ldots, n$ and $i, k \in \mathbb{Z}^+$. The commutation surfaces and equilibrium points introduced by the function are located when the corresponding domain $\mathcal{D}_i$ is being visited by the trajectory. It is important to mention that if α corresponds to the distance between two continuous equilibrium points given a value of c, then, each

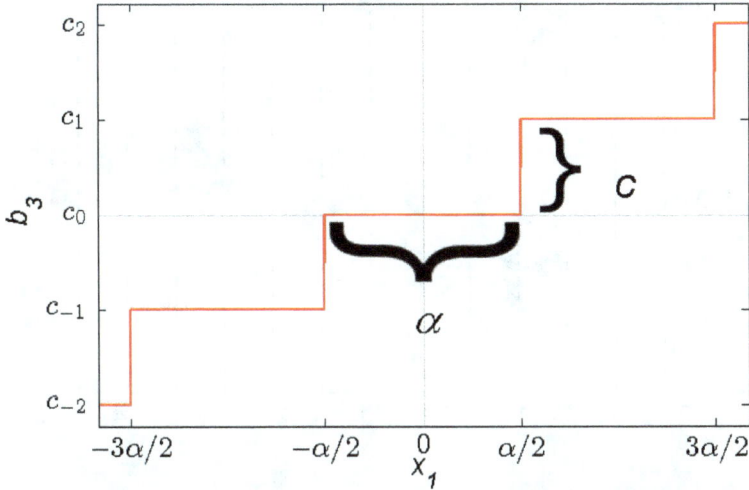

Fig. 2.9 Function b_3 given by (2.37). The distance between each step is given by α.

equilibrium point is located exactly at the middle of two consecutive commutation surfaces. Besides, the equilibria of the system and the domains are located from $(-\infty, \infty)$ due to the fact that the function given by (2.37) is not bounded as the commutation law in (2.33) or (2.35) are. This can be understand in the following way, if the system increases the size of its scroll, eventually it crosses to the next or previous $\mathcal{D}_i$ changing the value of c and increasing the number of scrolls in the system. This number of scrolls continues increasing according with $t \to \infty$.

In Fig. 2.10 a projection of the trajectory of the system (2.1) given by Eqs. (2.6) (2.32) and (2.37) into the plane (x_1, x_2) is displayed, presenting each commutation surface and equilibrium point between $-3 < x_1 < 5$ generated from the signal (2.37) considering $c = 6.3$, $\alpha = 0.6$ and $\mathbf{X}_0 = (-0.1, 0, 0)^T$. Notice that the oscillating behavior is similar as the one based on the commutation surfaces given by Eq. (2.35), but 9 scrolls are presented in this case for 50,000 iterations using the fourth order Runge–Kutta method. If the number of iterations are increased so will the number of scrolls.

The relationship between these two parameters satisfies $c/\alpha = 10.5$ which is equal to the entry a_{31} of the matrix $\mathbf{A}$, resulting in equilibrium points equally located between the commutation surfaces.

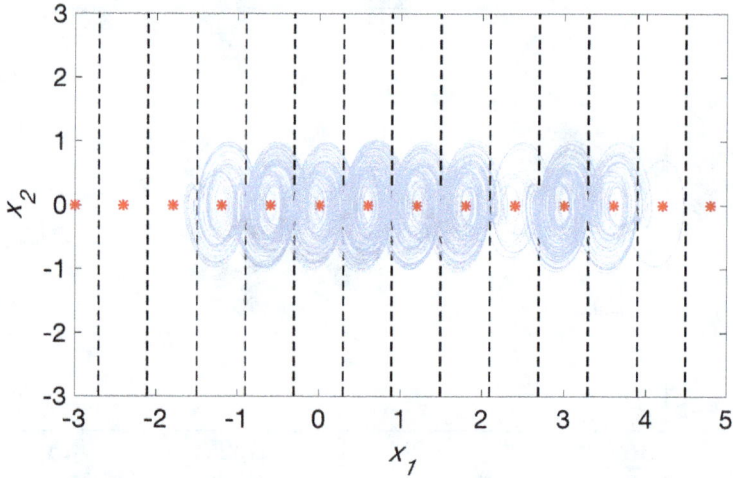

Fig. 2.10   Projection of the trajectory of the system (2.1) given by Eqs. (2.6) (2.32) and (2.37) onto the (x_1, x_2) plane with $c = 6.3$ and $\alpha = 0.6$. Marked with red asterisk the equilibria of the system, and with gray line the commutation surfaces generated by the function (2.37).

By using this approach, the equilibrium points and the position of the commutation surfaces are maintained as the obtained with Eq. (2.35), which is given as

$$b_3(x_1) = \begin{cases} c_i, & \text{if } x_1 \geq x_{i_{cs}}; \\ c * round(x_1/\alpha), & \text{if } x_{-i_{cs}} < x_1 < x_{i_{cs}}; \\ c_{-i}, & \text{if } x_1 \leq x_{i_{cs}}; \end{cases} \qquad (2.38)$$

where $i = 2$. Notice that $i \in \mathbb{Z}$ determines the number of scrolls in the attractor as $2 * i + 1$.

2.3.2   *Attractors generated by UDS type II*

The mechanism of generation of chaotic attractors based on this class of systems is due to the stable and unstable manifolds related with the design of A and B; this is, by considering two domains S_i and S_{i+1}, and the commutation surface Σ_τ between them. When the trajectory $\phi(x_0)_t$ reaches to the commutation surface Σ_τ due to the unstable manifold and initial condition $x_0 \in S_i$, and crosses the domain S_{i+1}, and the trajectory approaches the point of equilibrium due to the stable manifold, but again scapes from this domain due to the unstable manifold, being the trajectory $\phi(x_0)_t$ trapped forming the chaotic attractor.

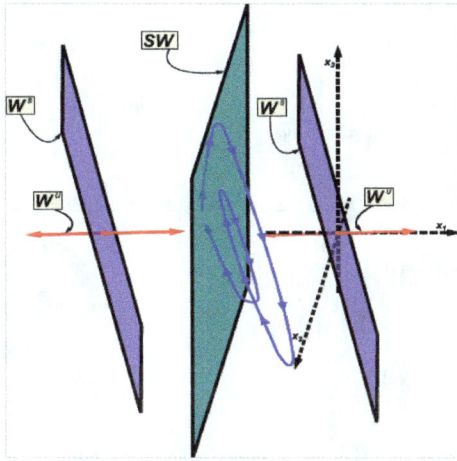

Fig. 2.11 System flow around SW.

The above result is illustrated by Fig. 2.11. Notice that the trajectory of the system oscillating around the unstable manifold W^u escapes from the domain S_i. This occurs near the unstable manifold $E^u \subset S_i$ where it crosses the commutation surface and it is rejected by the unstable manifold $W^u \subset S_{i+1}$ towards the equilibrium point x_i^* in the domain S_i. The process is repeated in the inverse way forming the attractor. For which, the set of all the initial conditions, in the phase space, whose corresponding trajectories converge to an attractor are defined as basin of attraction, Ω.

Consider the following linear operators:

$$A = \begin{pmatrix} 0 & 1 & 0 \\ 0 & 0 & 1 \\ -\alpha & -\beta & -\gamma \end{pmatrix}, \quad B(x_1) = \begin{pmatrix} 0 \\ 0 \\ \sigma(x_1) \end{pmatrix}; \qquad (2.39)$$

we consider the following set of parameters $\{a = -0.3494, b = 5.9469, c = -0.0988\}$. Then, according to the results of the Proposition 2.1, $\alpha = -0.6$, $\beta = 6$ and $\gamma = 0.6$. With this selection of parameters the eigenvalues of A are $\lambda_1 = 0.0989$ and $\lambda_{2,3} = -0.3494\pm2.4386i$, which according to Definition 2.5, the system is an UDS *Type II*. In particular, for this second example we define the following switching law:

$$\sigma^*(\chi) = \begin{cases} 0, \text{ if } x \in S_1 = \{x \in \mathbf{R}^3 : -1 \leq x_1\}; \\ 7, \text{ if } x \in S_2 = \{x \in \mathbf{R}^3 : x_1 < -1\}. \end{cases} \qquad (2.40)$$

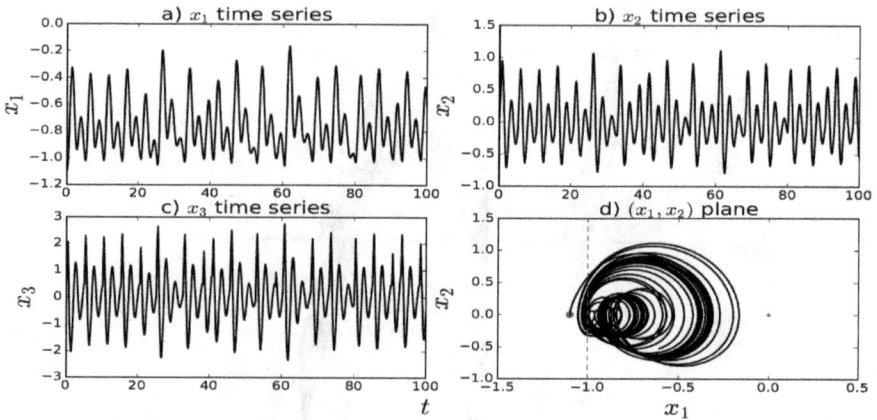

Fig. 2.12   (a) x_1 time series. (b) x_2 time series. (c) x_3 time series. (d) Projection of the attractor onto the (x_1, x_2) plane based on UDS *Type II*, with control parameters $a = -0.3494$, $b = 5.9469$ and $c = -0.0988$; and switching law (2.40). The dashed lines mark the division between the switching surfaces and the red dot indicates the initial position at $x_0 = (-1.1, 0, 0)^\top$.

Then, the equilibria for this system are located at $x_1^* = (0, 0, 0)^T$ and $x_2^* = (-11.6667, 0, 0)^T$. The unstable manifolds E_1^u and E_2^u lead the trajectory $\phi_t(x_0)$ toward the switching surface $\Sigma_1 = \{x \in \mathbb{R}^3 | x_1 = -1\}$ and $\chi_0 \in \Omega$. The basin of attraction Ω is between the stable manifolds E_1^s and E_2^s. In Fig. 2.12 we illustrate the dynamics of a switching system based on UDS-*Type II* for the initial conditions $x_0 = (-1.1, 0, 0)^\top$.

2.3.3   *Scroll attractors generated by the two types of UDS*

Now it is presented a generalized theory which is capable of explaining different approaches to construct double-scroll strange attractors based on the use of two different types of saddle-focus equilibrium points (i.e., taking into account different stability of each equilibrium point) and step functions in R^3.

It is important emphasize that the double-scroll attractor displayed by Chua system has three equilibrium points and does not display a scroll around the equilibrium point at the origin. This is because at least two saddle-focus equilibrium points of the same class are needed with local unstable manifolds of dimension two (the complex conjugate roots of

$\mathcal{P}(Df(x^*))$ have positive real part, where $\mathcal{P}(\cdot)$ denotes the characteristic polynomial of an operator and $Df(x^*)$ denotes the Jacobian matrix of f evaluated at the equilibrium pint x^*) and the equilibrium point at the origin has a stable manifold of dimension two.

There exist systems with chaotic attractors whose main behavioral mechanism has been explained through the presence of homoclinic orbits or heteroclinic loops (Section 2.2) for some selection of parameters.

A *homoclinic orbit* is defined as a bounded dynamical trajectory of the system that is *doubly asymptotic* to an equilibrium point. A *heteroclinic orbit* is similar except that there are two distinct saddle foci being connected, one corresponding to the forward asymptotic time limit. A *heteroclinic loop* is formed by the union of two or more heteroclinic orbits.

For the proposed class of systems, a combination of two saddle-focus equilibrium points of different class are considered. The idea behind the construction is based on the presence of a heteroclinic loop, however the analysis of that loop is out of the scope of this work.

Consider the piecewise linear (PWL) system with an associated vector field of the form:

$$\dot{x} = \begin{cases} A_1 x + B_1, & \text{if } s \leq 0; \\ A_2 x + B_2, & \text{if } s > 0; \end{cases} \tag{2.41}$$

where $x = [x_1, x_2, x_3]^T \in \mathbb{R}^3$ is the state vector, $B_1, B_2 \in \mathbb{R}^3$ are real constant vectors, $s = x_1 + x_2$ defines the switching plane $S = \{x \in \mathbb{R}^3 : x_1 + x_2 = 0\}$ and $A_1, A_2 \in \mathbb{R}^{3 \times 3}$ are linear operators whose associated polynomials $\mathcal{P}(A_1)$ and $\mathcal{P}(A_2)$ present a non-zero real root and two complex conjugate roots whose real part has the opposite sign of the real one. The real root of $\mathcal{P}(A_1)$ has also the opposite sign of the real root of $\mathcal{P}(A_2)$. Thus each subsystem of the form $\dot{x} = A_i x + B_i$ has a saddle-focus equilibrium point of different class. With an appropriate selection of subsystems it is possible to generate a chaotic double scroll attractor.

In Fig. 2.13 there is an illustration of the mechanism responsible for the attractor existence. The system is piecewise linear, so there is a subsystem assigned to each subset of the phase space, for this case the phase space is divided by a plane S that generates two subsets. In one of the subsets there is a saddle-focus equilibrium point x_A^* whose local stable manifold $W_{x_A^*}^S$ is one dimensional and its local unstable manifold $W_{x_A^*}^U$ is of dimension two.

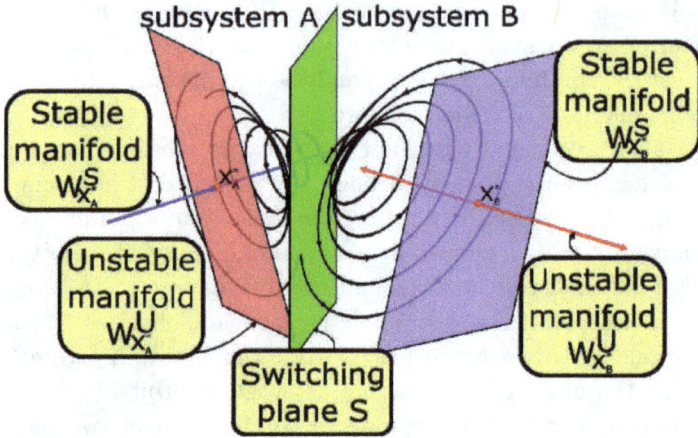

Fig. 2.13   Generation of the attractor by a PWL system.

Since $W_{x_A^*}^U$ is a plane not parallel to the switching plane S any trajectory in that subset close to x_A^* will eventually go through the switching plane in a point close to or in $S \cap W_{x_A^*}^U$. In the other subset of the phase space there is the equilibrium point x_B^* whose local stable manifold $W_{x_B^*}^S$ is two-dimensional and its local unstable manifold $W_{x_B^*}^U$ is of dimension one. The manifold $W_{x_B^*}^U = \text{span}\{u\}$ for a vector $u \in \mathbb{R}^3$ and $u \notin S$, furthermore $W_{x_B^*}^S$ divides that subset of the phase space in two subsets, in one of them the unstable manifold directs the flow against S. Then any trajectory crossing the plane S in that subset will return to S. The orientations of the local manifolds will determine the geometry of the attractor once the trajectories are bounded in this oscillating process around S.

In Fig. 2.14 three qualitative trajectories are shown. The initial condition for the trajectories $x_0 \in W_{x_A^*}^U$ is a point close to the equilibrium point x_A^*. The first trajectory shown in Fig. 2.14(a) is the case where the trajectory goes through the point $x_c \in W_{x_A^*}^U \cap S$ and then enters into a region where the orbit diverges not forming an attractor. A second case shown in Fig. 2.14(b), is when $W_{x_A^*}^U \cap cl(W_{x_B^*}^S) \neq \emptyset$ where $cl(\cdot)$ denotes the closure, the trajectory goes through the point $x_c \in W_{x_A^*}^U \cap cl(W_{x_B^*}^S)$ and tends to x_B^* as $t \to \infty$, thus an heteroclinic loop is formed. In the same way $cl(W_{x_B^*}^U) \cap W_{x_A^*}^S \neq \emptyset$ and then any trajectory starting in $W_{x_B^*}^U$ goes to x_A^* as $t \to \infty$ which form a second heteroclinic loop and a heteroclinic orbit is completed. Note that if $W_{x_A^*}^U \cap cl(W_{x_B^*}^U) \cap S \neq \emptyset$ and $W_{x_A^*}^S \cap cl(W_{x_B^*}^U) \cap S \neq \emptyset$ is guaranteed then there exists a heteroclinic orbit. A third case is shown in Fig. 2.14(c) where the trajectory goes through $W^U_{x^*_A} \cap S$ and enters into a region where the equilibrium point x^*_B repels the trajectory against S which traps the trajectory between the two saddle equilibrium points and generates a chaotic attractor.

In Fig. 2.15 the three trajectories cases aforementioned, respectively, are presented from another perspective view and the two-dimensional manifolds are represented as a line that corresponds to a vector of the two-dimensional manifold. This representation allows us to easily see the role of the manifolds in the behavior of the system close to the equilibria.

The interest is to form a chaotic attractor as is depicted in third case aforementioned. The matrix A_1 has been choosen in controllable canonical form due to its simplicity, the matrix A_2 presents an apparent more complicated form, however, its form provide a useful location of the manifolds and the eigenvalues for the proposed construction, once that the matrix A_i are established and according to the their manifolds the location of the equilibrium points are selected based on the aforementioned idea, then the vectors B_1 and B_2 are obtained by $-A_i x^* = B_i$. To illustrate a construction that exhibits this behavior as the third case consider the particular system of the form (2.41) with:

$$A_1 = \begin{bmatrix} 0 & 1 & 0 \\ 0 & 0 & 1 \\ 0.313 & -6.25 & -0.15 \end{bmatrix},$$

$$A_2 = \begin{bmatrix} 0.9364 & 0.2609 & 1.3045 \\ -0.8548 & -0.1524 & -0.2619 \\ -5.1564 & -0.1168 & -0.6840 \end{bmatrix},$$

$$(2.42)$$

$$B_1 = \begin{bmatrix} 1 \\ -1 \\ -5.474 \end{bmatrix}, \quad B_2 = \begin{bmatrix} -2.3306 \\ 1.0869 \\ 4.9756 \end{bmatrix}. \qquad (2.43)$$

The roots of $\mathcal{P}(A_1)$ are $\lambda_1 = 0.05, \lambda_{2,3} = -0.1 \pm 2.5i$ and the roots of $\mathcal{P}(A_2)$ are $\lambda_1 = -0.1, \lambda_{2,3} = 0.1 \pm 2.5i$. The resulting double-scroll attractor is shown in Fig. 2.19. It is worth mentioning that the double-scroll attractor generated by our Chua-like system differentiates from other Chua-like attractors where in order to get two scrolls, there is a need of at least two saddle-focus equilibrium points of the same class with local unstable manifolds of dimension two (the complex conjugate roots of $\mathcal{P}(Df(x^*))$ have positive real part). Figure 2.17 shows the time series of the system,

(a)

(b)

(c)

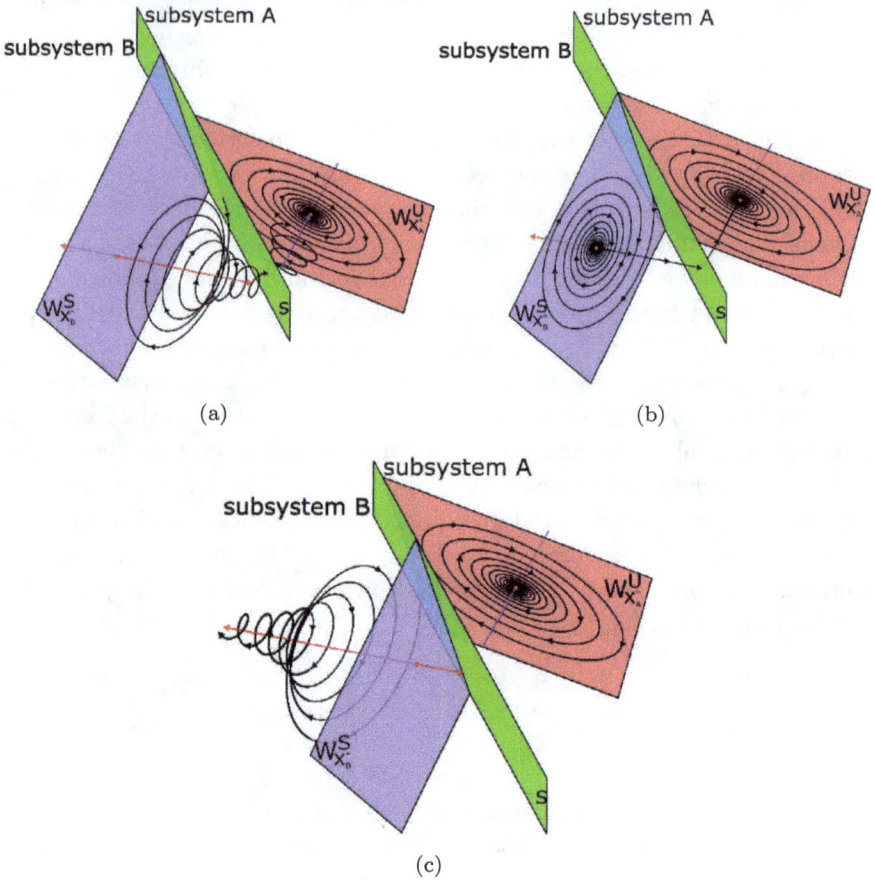

Fig. 2.14   Three qualitative trajectories in the PWL system with two different saddle focus equilibrium points. (a) Trajectory is trapped between the two equilibrium points. (b) Heteroclinic orbits are generated. (c) Trajectory escapes and gets away the surface S.

which are a mixed underdamped and increasing oscillations signals which are in concordance with the two types of equilibrium points.

For a second approach to generate a double-scroll chaotic attractor via UDS-I and UDS-II, we consider a PWL system defined by two domains D_1 and D_2 as follows:

$$\dot{\mathbf{x}} = \begin{cases} A_1\mathbf{x} + B_1, & \text{if } x_1 + x_3 < 0; \\ A_2\mathbf{x} + B_2, & \text{if } x_1 + x_3 \geq 0; \end{cases} \tag{2.44}$$

where $\mathbf{x} = (x_1, x_2, x_3)^T$ is the state vector, $A_1\mathbf{x} + B_1$ and $A_2\mathbf{x} + B_2$ are UDS II and I systems, respectively. The linear operators A_1 and A_2 are

(a)

(b)

(c)

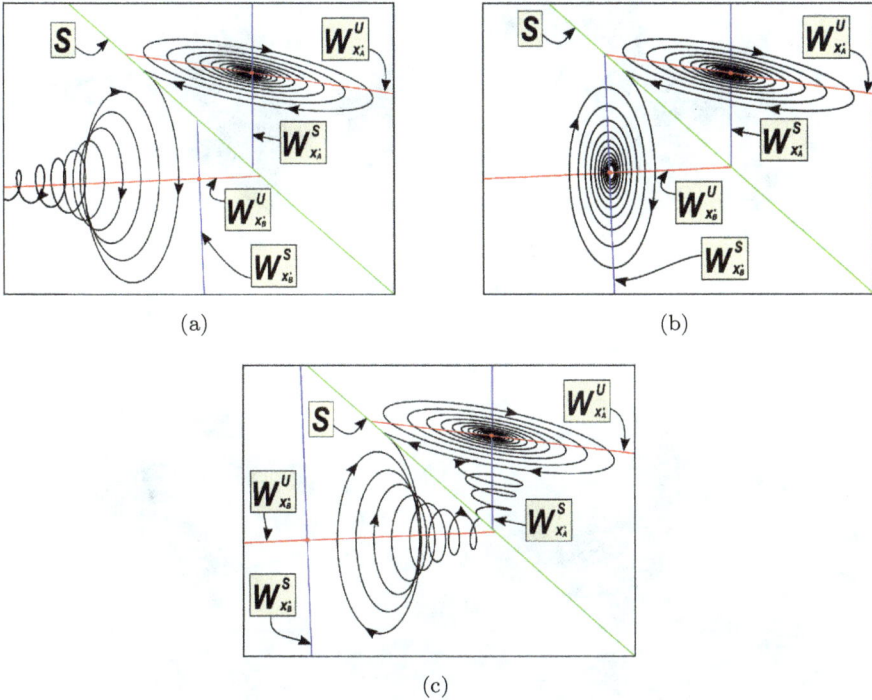

Fig. 2.15 Three qualitative trajectories of Fig. 2.14 seen from another angle and where the two-dimensional manifolds are represented only by a line that correspond to a vector of the two-dimensional manifold.

given in Jordan block form:

$$A_1 = \begin{pmatrix} -\sigma_1 & \omega & 0 \\ -\omega & -\sigma_1 & 0 \\ 0 & 0 & \gamma_1 \end{pmatrix}, \quad A_2 = \begin{pmatrix} -\gamma_2 & 0 & 0 \\ 0 & \sigma_2 & -\omega \\ 0 & \omega & \sigma_2 \end{pmatrix}, \quad (2.45)$$

with $0 < \sigma_1, \sigma_2, \omega, \gamma_1, \gamma_2 \in \mathbb{R}$ such that $\sigma_1 + \gamma_1 = \sigma_2 + \gamma_2$. $B_1, B_2 \in \mathbb{R}^3$ are constant vectors. It is possible to find B_1 and B_2 such that the existence of a heteroclinic loop is guaranteed since A_1 and A_2 are invertible matrices. The equilibrium points $\mathbf{x}_1^*$ and $\mathbf{x}_2^*$ are given by $\mathbf{x}_1^* = -A_1^{-1}B_1$ and $\mathbf{x}_2^* = -A_2^{-1}B_2$. Then it is always possible to locate the equilibrium points satisfying $\mathbf{x}_1^* = -\mathbf{x}_2^*$ with $\mathbf{x}_1^* = (-k, 0, -k)^T$ and $0 < k \in \mathbb{R}$. When the equilibria satisfy this last condition, the local linear manifolds $W_{\mathbf{x}_1^*}^U = \text{span}\{(0,0,1)^T\} - (k,0,k)^T$ and $W_{\mathbf{x}_2^*}^S = \text{span}\{(1,0,0)^T\} + (0,0,k)^T$ intersect at the point $(-k, 0, k)^T$. This point of intersection belongs to the switching plane $S = \{\mathbf{x} \in \mathbb{R}^3 : x_1 + x_3 = 0\}$ and thus a heteroclinic orbit is conformed.

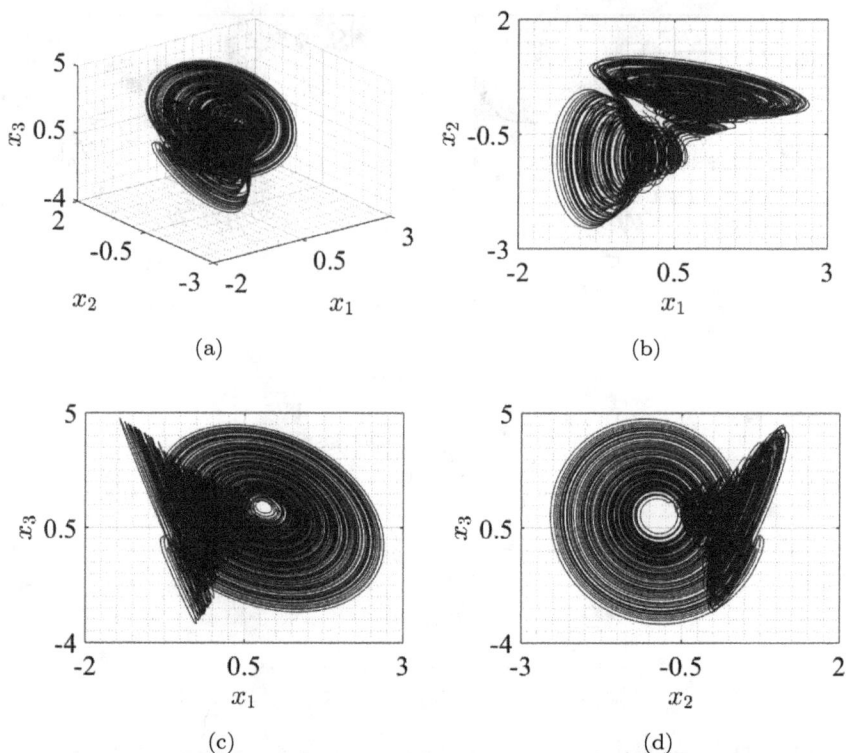

Fig. 2.16   Attractor of the system (2.41) with A_1, A_2, B_1 and B_2 for the initial condition $(0, 0, 0)$ ($t = 1000s$) in (a) the space (x_1, x_2, x_3) and its projections onto the planes: (b) (x_1, x_2), (c) (x_1, x_3) and (d) (x_2, x_3).

The local manifolds $W_{\mathbf{x}_1^*}^S$ and $W_{\mathbf{x}_2^*}^U$ are defined by the planes $\{\mathbf{x} \in \mathbb{R}^3 : x_3 = -k\}$ and $\{\mathbf{x} \in \mathbb{R}^3 : x_1 = k\}$, respectively. The intersection of the two planes is the subset of the space given by $\text{span}\{(0, 1, 0)^T\} + (k, 0, -k)^T$ which is indeed part of the switching plane. There is a point in the switching plane that generates a heteroclinic orbit because its trajectory goes to $\mathbf{x}_2^*$ when $t \to -\infty$ and converges to $\mathbf{x}_1^*$ when $t \to \infty$. There is a second heteroclinic orbit and thus a heteroclinic loop is conformed. The idea of the proposed construction is to start with a system given by (2.44) that presents a heteroclinic loop, and in this way it is possible to obtain a double-scroll chaotic attractor.

The space is partitioned by the switching plane S in $D_1 = \{\mathbf{x} \in \mathbb{R}^3 : x_1 + x_3 < 0\}$ and $D_2 = \{\mathbf{x} \in \mathbb{R}^3 : x_1 + x_3 \geq 0\}$. The subset D_2 contains

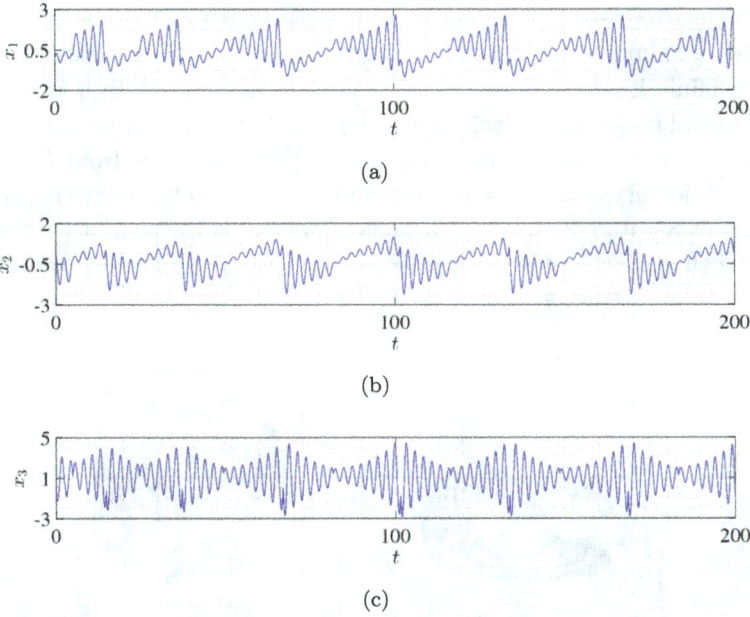

Fig. 2.17   Time series of the attractor shown in Fig. 2.19. In (a) x_1 state, (b) x_2 state and (c) x_3 state.

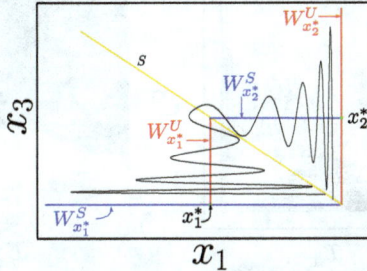

Fig. 2.18   Illustration of a trajectory of the system given (2.44) projected onto the plane (x_1, x_3).

the saddle-focus equilibrium point $\mathbf{x}_2^*$ with a one-dimensional stable manifold $W_{\mathbf{x}_2^*}^S$ and a two-dimensional unstable manifold $W_{\mathbf{x}_2^*}^U$. Since $W_{\mathbf{x}_2^*}^U$ is a plane not parallel to the switching plane S, any trajectory in a subset close to $\mathbf{x}_2^*$ will eventually go through the switching plane in a region close to $S \cap W_{\mathbf{x}_2^*}^S$. The other subset D_1 of the state space contains the equilibrium point $\mathbf{x}_1^*$ with a two-dimensional stable manifold $W_{\mathbf{x}_1^*}^S$ and one-dimensional unstable manifold $W_{\mathbf{x}_1^*}^U$. An illustration of the manifolds for the case when a heteroclinic loop exists is presented in the Fig. 2.18. Also it is shown a hypothetical trajectory (black curve) that starts in D_1 near to the planes S and $W_{\mathbf{x}_1^*}^s$, and goes to a neighborhood of $W_{\mathbf{x}_1^*}^U \cap S$ to scape from D_1 to D_2. So, the trajectory crosses the plane S and enters into D_2, where the equilibrium point $\mathbf{x}_2^*$ attracts and at the same time repels the trajectory towards S. Thus the trajectory is trapped between the two saddle-foci equilibrium points and an attractor is generated.

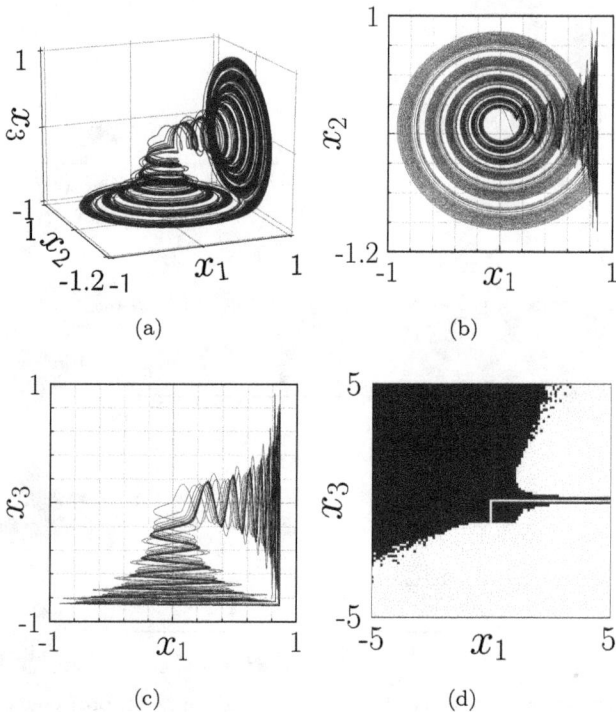

Fig. 2.19   Attractor of the system given by (2.44), (2.45), and (2.46) with $\gamma_1 = 0.2$, $\gamma_2 = 0.25$, $\sigma_1 = 0.2$, $\sigma_2 = 0.15$ and $\omega = 4$, simulated for the initial condition $\mathbf{x}(0) = (0.8, 0.2, -0.8)^T$ and $t \in [0, 1000]$ in (a) the space (x_1, x_2, x_3) and its projections onto the planes: (b) (x_1, x_2) and (c) (x_1, x_3). In (d) the estimated basin of attraction in the plane $\{\mathbf{x} \in \mathbb{R}^3 : x_2 = 0\}$ is shown in blue.

As particular case the parameters have been chosen as $\gamma_1 = 0.2$ $\gamma_2 = 0.25$ $\sigma_1 = 0.2$, $\sigma_2 = 0.15$ and $\omega = 4$. The two equilibrium points are set symmetrically separated from the switching plane S and are given by $\mathbf{x}_1^* = (0, 0, -1)^T$ and $\mathbf{x}_2^* = (1, 0, 0)^T$. Thus $B_1 = -A\mathbf{x}_1^*$ and $B_2 = -A\mathbf{x}_2^*$. It is easy to see that the intersection of $W_{\mathbf{x}_1^*}^U$ and the closure of $W_{\mathbf{x}_2^*}^S$ is different to the empty set as well as the intersection on the closure of $W_{\mathbf{x}_2^*}^U$ and $W_{\mathbf{x}_1^*}^S$. Therefore the existence of the heteroclinic loop is evident. Now in order to obtain a better looking attractor we "perturb" the system to change the equilibrium point $\mathbf{x}_2^*$ to $\mathbf{x}_2^* = (0.9, 0, 0)^T$ and keep only one heterclinic orbit. Thus B_1 and B_2 are given by:

$$B_1 = \begin{pmatrix} 0 \\ 0 \\ 0.2 \end{pmatrix}, \quad B_2 = \begin{pmatrix} 0.225 \\ 0 \\ 0 \end{pmatrix}. \tag{2.46}$$

The system attractor is shown in the Fig. 2.19. The estimated basin of attraction in the plane $\{\mathbf{x} \in \mathbb{R}^3 : x_2 = 0\}$ is shown in the Fig. 2.19(d).

In order to verify the chaotic behavior of the attractor the Maximum Lyapunov Exponent (MLE) is calculated, the obtained value is MLE=0.166037.

2.3.4 *Quadruple-scroll attractor via UDS-I and UDS-II*

By means of increasing the number of domains in the partition of the phase space. Our starting point is the PWL system previously defined in Section 2.3.3. Now, the phase space is partitioned in four domains D_i, with $i = 1, \dots 4$, by considering the previous switching plane $S_1 = \{\mathbf{x} \in \mathbb{R}^3 : x_1 + x_3 = 0\}$ and a new one given by $S_2 = \{\mathbf{x} \in \mathbb{R}^3 : x_1 - x_3 = 0\}$. Each domain D_i has a stable manifold or an unstable manifold given by planes, this allows the generation of an attractor at the other side of the switching surface S_2 and around switching surface S_1

So the PWL system is given as follows:

$$\dot{\mathbf{x}} = \begin{cases} A_1\mathbf{x} + B_1, & \text{if } x_1 + x_3 < 0 \text{ and } -x_1 + x_3 < 0; \\ A_2\mathbf{x} + B_2, & \text{if } x_1 + x_3 \geq 0 \text{ and } -x_1 + x_3 \leq 0; \\ A_2\mathbf{x} + B_4, & \text{if } x_1 + x_3 > 0 \text{ and } -x_1 + x_3 > 0; \\ A_1\mathbf{x} + B_3, & \text{otherwise}; \end{cases} \tag{2.47}$$

with A_1 and A_2 given by (2.45) and B_i with $i = 1, \dots, 4$ given by:

$$B_1 = -B_4 = \begin{pmatrix} 0.1 \\ 2 \\ 0.3 \end{pmatrix} \text{ and } B_2 = -B_3 = \begin{pmatrix} 0.35 \\ -2 \\ 0.075 \end{pmatrix}, \tag{2.48}$$

presents four scrolls in the attractor, i.e., a quadruple-scroll attractor with square shape which will be called *square chaotic attractor*. The two hetero-clinic orbits collide at $(0, 0, 0)^T$ and then the unstable manifolds of $\mathbf{x}_1^*$ and $\mathbf{x}_4^*$ are joined to the stable manifold of $\mathbf{x}_2^*$. So no matter where an initial condition starts if it belongs to the basin of attraction, then the trajectory converges to the attractor around the intersection of the switching planes, otherwise the trajectory grows to infinite.

Figure 2.20 shows the attractor of the system given by (2.45), (2.47), and (2.48) with $\gamma_1 = 0.2$ $\gamma_2 = 0.25$ $\sigma 1 = 0.2$, $\sigma_2 = 0.15$, $\omega = 4$, and $k = 0$ in the space (x_1, x_2, x_3) and its projections onto the planes: (b) (x_1, x_2), (c) (x_1, x_3), and (d) (x_2, x_3). The chaotic behavior of the four scroll attractor was verified by the maximum Lyapunov exponent MLE=0.162278. The test for Chaos 1-0 is also used to verify the chaotic behavior, the result is $K = 0.9711$ which is an indicative of chaos.

2.4   Approximation of continuous systems to PWL systems

Linear and linear affine systems has been widely used to construct Piece-wise linear systems that exhibits at least one chaotic attractor. PWL systems are in general easier to study that continuous nonlinear systems. Thus, an interesting topic of study is the approximation of known systems to a PWL version that preserves some characteristics as the existence of the chaotic attractor. This idea have been addressed in some works [Petrzela *et al.* (2003); Amaral *et al.* (2006)].

According to Amaral *et al.* (2006), the steps that can be followed are:

- Define the number and location of fixed points around which the dynamics will be organized.
- Evaluate the Jacobian matrix at each of the fixed points of the focus type defined in the previous step.
- Determine a switching surface based on topological guidelines and/or the relative organization of the fixed point.
- Simulate the piece-wise affine model and compare the resulting attractor by means of topological analysis. The position of the switching surface determined in the previous step can be used for fine tuning.

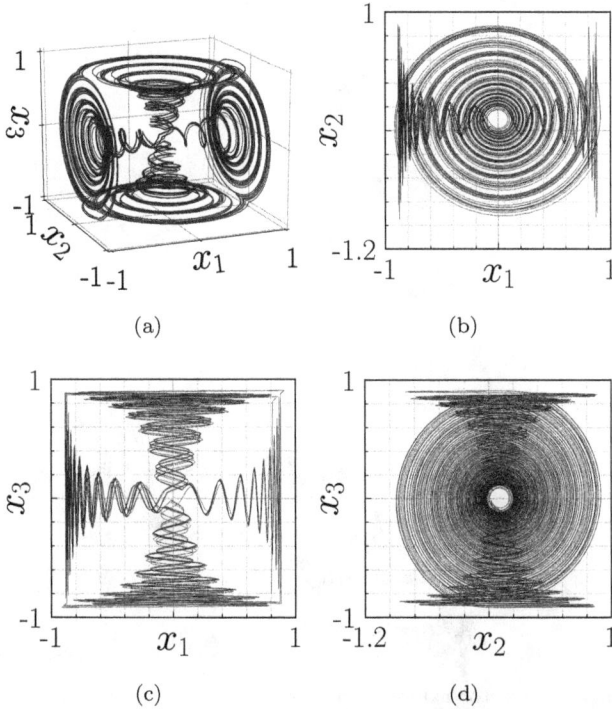

Fig. 2.20 Attractor of the system given by (2.45), (2.47), and (2.48) with $\gamma_1 = 0.2$ $\gamma_2 = 0.25$ $\sigma = 0.2$, $\sigma_2 = 0.15$, $\omega = 4$, and $k = 0$ simulated for the initial condition $\mathbf{x}(0) = (0.8, 0.2, -0.8)^T$ and $t \in [0, 1000]$ in (a) the space (x_1, x_2, x_3) and its projections onto the planes: (b) (x_1, x_2), (c) (x_1, x_3), and (d) (x_2, x_3).

To illustrate this idea let us consider the example given in [Amaral *et al.* (2006)] about the Rössler system:

$$\begin{aligned}
\dot{x} &= -y - z \\
\dot{y} &= x + ay \\
\dot{z} &= b + z(x - c)
\end{aligned} \tag{2.49}$$

with $a = 0.2$, $b = 0.2$ and $c = 5.7$. The chaotic attractor is shown in the Fig. 2.21. Thus, in order to approximate the attractor the equilibria is found. The system (2.49) has two equilibrium points:

$$x_{eq_1} = \begin{pmatrix} \frac{c + \sqrt{c^2 - 4ab}}{2} \\ \frac{-c - \sqrt{c^2 - 4ab}}{2a} \\ \frac{c + \sqrt{c^2 - 4ab}}{2a} \end{pmatrix}, \tag{2.50}$$

(a)   (b)

(c)   (d)

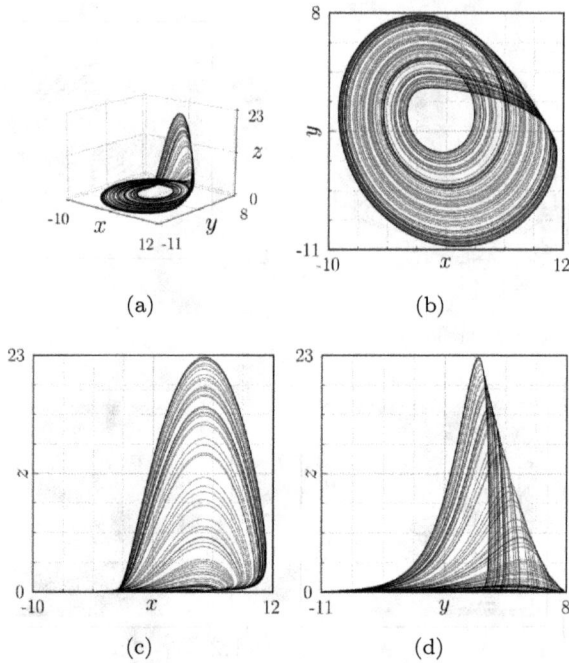

Fig. 2.21 Chaotic attractor exhibited by the Rössler system in (a) $x - y - z$, and its projections onto the planes (b) $x - y$, (c) $x - z$ and (d) $y - z$.

$$x_{eq_2} = \begin{pmatrix} \frac{c - \sqrt{c^2 - 4ab}}{2} \\ \frac{-c + \sqrt{c^2 - 4ab}}{2a} \\ \frac{c - \sqrt{c^2 - 4ab}}{2a} \end{pmatrix}. \tag{2.51}$$

Then the Jacobean matrix is evaluated at each of the fixed points

$$\begin{pmatrix} 0 & -1 & -1 \\ 1 & a & 0 \\ z_{eq_i} & 0 & x_{eq_i} - c \end{pmatrix} \tag{2.52}$$

for x_{eq_1} the eigenvalues are:

$$\begin{aligned} \lambda_1 &= -4.596071039253413e - 6 - 5.428025931106868i \\ \lambda_2 &= -4.596071039253413e - 6 + 5.428025931106868i \\ \lambda_3 &= 0.19298298730797747 \end{aligned} \tag{2.53}$$

for x_{eq_2} the eigenvalues are:

$$\lambda_1 = 0.09700085608848114 - 0.9951934910299535i$$
$$\lambda_2 = 0.09700085608848114 + 0.9951934910299535i \qquad (2.54)$$
$$\lambda_3 = -5.686975507342861.$$

Both equilibrium points are of type saddle focus. Then according to [Amaral *et al.* (2006)] the system can be written as

$$\begin{pmatrix} \dot{x} \\ \dot{y} \\ \dot{z} \end{pmatrix} = f[s_1(X)] \begin{pmatrix} 0 & -1 & -1 \\ 1 & a & 0 \\ z_{eq_1} & 0 & x_{eq_1} - c \end{pmatrix} \begin{pmatrix} x - x_{eq_1} \\ y - y_{eq_1} \\ z - z_{eq_1} \end{pmatrix}$$

$$- f[s_2(X)] \begin{pmatrix} 0 & -1 & -1 \\ 1 & a & 0 \\ z_{eq_2} & 0 & x_{eq_2} - c \end{pmatrix} \begin{pmatrix} x - x_{eq_2} \\ y - y_{eq_2} \\ z - z_{eq_2} \end{pmatrix} \qquad (2.55)$$

where

$$f_i[s(X)] = \begin{array}{l} 1 \text{ if } s(X) \geq 0 \\ 0 \text{ if } s(X) < 0 \end{array} \qquad (2.56)$$

with $s_1(X) = -0.95y^2 - 3y - 6 + x$ and $s_2(X) = -s_1(X)$. The resulting attractor is shown in the Fig. 2.22. Note that the switching surface can be selected different leading to a different shape in the attractor, as example consider $s_1(X) = x - 3.59$, the resulting attractor is shown in the Fig. 2.23.

This system has stationary, periodic, quasi-periodic, and chaotic attractors that depend on the value of the parameters (a, b, c) [Li *et al.* (2014)]. These attractors are connected by bifurcations, in particular by a Hopf bifurcation from the stationary to the periodic attractor and a period doubling cascade from the periodic to the chaotic behavior. The resulting chaotic attractor has a single lobe and is called spiral-type chaos for $a = 0.16$ $b = 0.1$ $c = 8.5$, manifesting itself mainly in irregular amplitudes of the oscillations.

Other approaches found in the literature includes the approximation on non-linearities as it has been done in [Li *et al.* (2014)]. In this work a linear feedback control from an additional variable (u) is introduced in the diffusionless Lorenz system to produce the following four-dimensional system:

$$\begin{aligned} \dot{x} &= y - x \\ \dot{y} &= -xz + u \\ \dot{z} &= xy - a \\ \dot{u} &= -by. \end{aligned} \qquad (2.57)$$

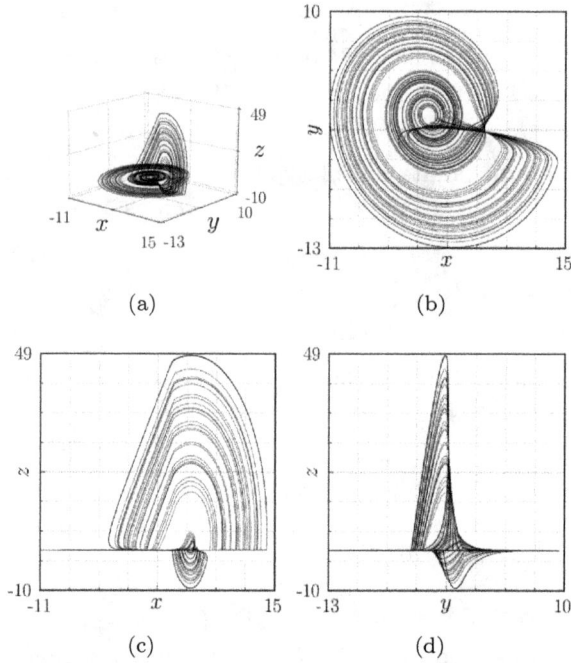

Fig. 2.22   Chaotic attractor of the system given by (2.55) and (2.55) with $s_1(X) = -0.95y^2 - 3y - 6 + x$ and $s_2(X) = -s_1(X)$ in (a) $x - y - z$, and its projections onto the planes (b) $x - y$, (c) $x - z$ and (d) $y - z$.

The system present a hyperchaotic attractor for $a = 2.7$ and $b = 0.44$ which is shown in the Fig. 2.24. According to this work non-linearities of the form x^2 can be replaced by $|x|$ and those of the form xy by $x\,\mathrm{sgn}(y)$ or $y\,\mathrm{sgn}(x)$.

For the system (2.57) the numerical test showed that the change of xy to $|x|$ preserved the hyperchaos and thus the system can be approximated as follows:

$$
\begin{aligned}
\dot{x} &= y - x \\
\dot{y} &= -z\,\mathrm{sgn}(x) + u \\
\dot{z} &= |x| - a \\
\dot{u} &= -by.
\end{aligned}
\tag{2.58}
$$

The modified given by (2.58) is shown in the Fig. 2.25.

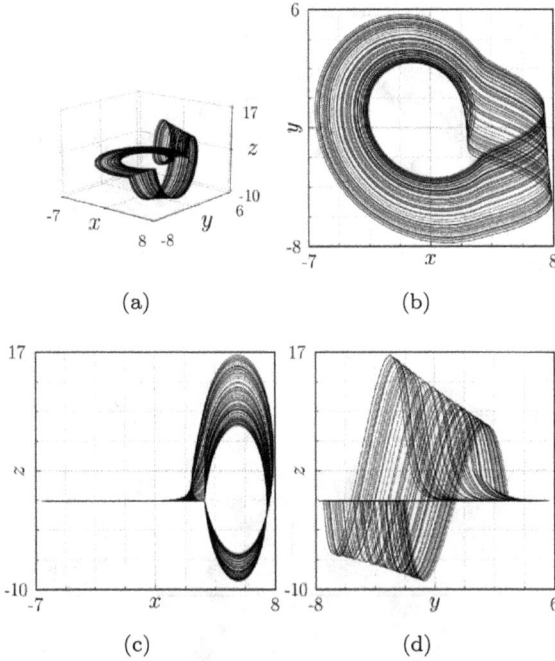

Fig. 2.23 Chaotic attractor of the system given by (2.55) and (2.55) with $s_1(X) = x - 3.59$ and $s_2(X) = -s_1(X)$ in (a) $x - y - z$, and its projections onto the planes (b) $x - y$, (c) $x - z$ and (d) $y - z$.

2.5 2D and 3D directional multi-scroll attractors

In Section 2.5, we expand the concept of 1D-grid to 2D and 3D-grid multi-scroll attractors based on UDS. We start by considering a PWL system given by

$$\dot{\mathbf{x}} = A\mathbf{x} + BF(\mathbf{x}), \qquad (2.59)$$

that is defined in a partition $\mathcal{P} = \{P_1, P_2, \ldots, P\eta\}$, with $1 < \eta \in \mathbb{Z}$. The location of the equilibria $\mathbf{x}_i^* = (x_{1i}^*, x_{2i}^*, x_{3i}^*)^T$, with $i = 1, \ldots, \eta$, of a PWL system (2.59) is determined by the linear operators A given by (2.18) and the matrix B and the vector F. Each entry of the vector F is controlled by a step functional defined in each atom p_i of the partition $\mathcal{P}$, with $i = 1, \ldots, \eta$. We consider the last three equilibria of 1D-grid multi-scroll attractor of Section 2.3 that are at $\mathbf{x}_1^* = (-\alpha, 0, 0)^T$, $\mathbf{x}_2^* = (\alpha, 0, 0)^T$, and $\mathbf{x}_3^* = (3\alpha, 0, 0)^T$. The linear operator A given by (2.18) considers $a = 0.2, b = 5$ and $c = -3$. Firstly, we discuss the generation of 2D-grid

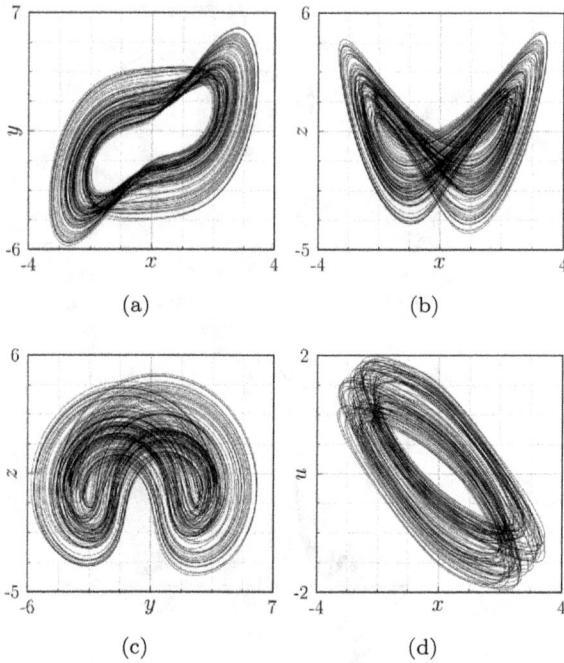

Fig. 2.24   Chaotic attractor of the system given by (2.57) in (a) $x-y$, (b) $x-z$, (c) $y-z$ and (d) $y-u$.

multi-scroll attractor by considering six atoms of a $\mathcal{P}$ partition containing a saddle equilibrium point in each of them. The six atoms are defined for $N_1 = (2, 0, 1)$, $N_2 = (0, 1, 0)$, $D_1 = 0$, $D_2 = 4$, and $D_3 = 0.7$ as follows

$$
\begin{aligned}
P_1 = {} & \{\mathbf{x} \in \mathbb{R}^3 : x_3 > 0,\ \mathbf{N_1} \cdot \mathbf{x} \leq -D_1,\ \mathbf{N_2} \cdot \mathbf{x} \leq -D_3\} \\
& \cup \{\mathbf{x} \in \mathbb{R}^3 : x_3 \leq 0,\ \mathbf{N_1} \cdot \mathbf{x} < -D_1,\ \mathbf{N_2} \cdot \mathbf{x} \leq -D_3\}, \\
P_2 = {} & \{\mathbf{x} \in \mathbb{R}^3 : x_3 > 0,\ \mathbf{N_1} \cdot \mathbf{x} > -D_1,\ \mathbf{N_1} \cdot \mathbf{x} \leq -D_2,\ \mathbf{N_2} \cdot \mathbf{x} \leq -D_3\} \\
& \cup \{\mathbf{x} \in \mathbb{R}^3 : x_3 \leq 0,\ \mathbf{N_1} \cdot \mathbf{x} \geq -D_1,\ \mathbf{N_1} \cdot \mathbf{x} < -D_2,\ \mathbf{N_2} \cdot \mathbf{x} \leq -D_3\} \\
P_3 = {} & \{\mathbf{x} \in \mathbb{R}^3 : x_3 > 0,\ \mathbf{N_1} \cdot \mathbf{x} > -D_2,\ \mathbf{N_2} \cdot \mathbf{x} \leq -D_3\} \\
& \cup \{\mathbf{x} \in \mathbb{R}^3 : x_3 \leq 0,\ \mathbf{N_1} \cdot \mathbf{x} \geq -D_2,\ \mathbf{N_2} \cdot \mathbf{x} \leq -D_3\}, \\
P_4 = {} & \{\mathbf{x} \in \mathbb{R}^3 : x_3 > 0,\ \mathbf{N_1} \cdot \mathbf{x} \leq -D_1,\ \mathbf{N_2} \cdot \mathbf{x} > -D_3\} \\
& \cup \{\mathbf{x} \in \mathbb{R}^3 : x_3 \leq 0,\ \mathbf{N_1} \cdot \mathbf{x} < -D_1,\ \mathbf{N_2} \cdot \mathbf{x} > -D_3\}, \\
P_5 = {} & \{\mathbf{x} \in \mathbb{R}^3 : x_3 > 0,\ \mathbf{N_1} \cdot \mathbf{x} > -D_1,\ \mathbf{N_1} \cdot \mathbf{x} \leq -D_2,\ \mathbf{N_2} \cdot \mathbf{x} > -D_3\} \\
& \cup \{\mathbf{x} \in \mathbb{R}^3 : x_3 \leq 0,\ \mathbf{N_1} \cdot \mathbf{x} \geq -D_1,\ \mathbf{N_1} \cdot \mathbf{x} < -D_2,\ \mathbf{N_2} \cdot \mathbf{x} > -D_3\} \\
P_6 = {} & \{\mathbf{x} \in \mathbb{R}^3 : x_3 > 0,\ \mathbf{N_1} \cdot \mathbf{x} > -D_2,\ \mathbf{N_2} \cdot \mathbf{x} > -D_3\} \\
& \cup \{\mathbf{x} \in \mathbb{R}^3 : x_3 \leq 0,\ \mathbf{N_1} \cdot \mathbf{x} \geq -D_2,\ \mathbf{N_2} \cdot \mathbf{x} > -D_3\}.
\end{aligned}
$$

$$(2.60)$$

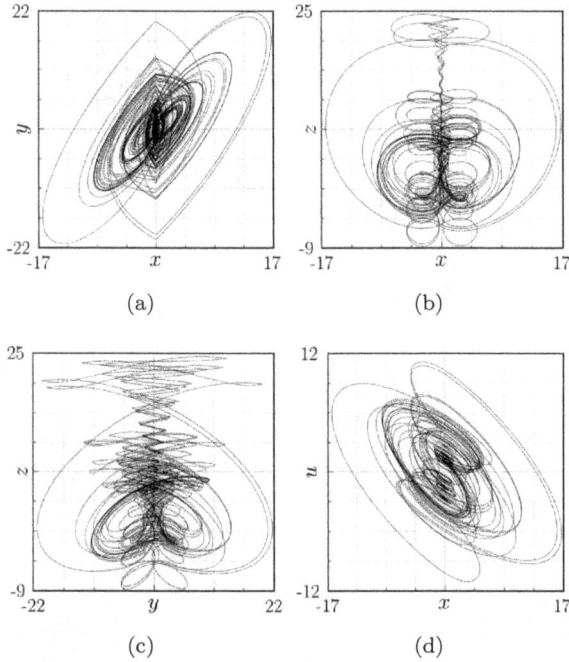

Fig. 2.25 Chaotic attractor of the system given by (2.57) in (a) $x-y$, (b) $x-z$, (c) $y-z$ and (d) $y-u$.

The switching surface are given between adjacent atoms, therefore three switching surfaces are defined as follows

$$\begin{aligned}
SW_{12} &= SW_{45} = \mathbf{N_1} \cdot \mathbf{x} = D_1 \\
SW_{23} &= SW_{56} = \mathbf{N_1} \cdot \mathbf{x} = D_2 \\
SW_{14} &= SW_{25} = SW_{36} = \mathbf{N_2} \cdot \mathbf{x} = D_3.
\end{aligned} \tag{2.61}$$

The constant matriz $B \in \mathbb{R}^{3 \times 2}$ is defined as follows:

$$B = \begin{pmatrix} -\dfrac{a+2c}{3} & -b \\ \dfrac{b}{3} & -a \\ \dfrac{a-c}{3} & b \end{pmatrix}, \tag{2.62}$$

and $\mathbf{F}(\mathbf{x}) = (f_1(\mathbf{x}), f_2(\mathbf{x}))^T$, where the functional f_1 is given by

$$f_1(\mathbf{x}) = \begin{cases} -\alpha, & \text{if } \mathbf{x} \in P_1 \cup P_4, \\ \alpha, & \text{if } \mathbf{x} \in P_2 \cup P_5, \\ 3\alpha & \text{if } \mathbf{x} \in P_3 \cup P_6. \end{cases} \tag{2.63}$$

And the functional f_2 is given by

$$f_2(\mathbf{x}) = \begin{cases} 0, & \text{if } \mathbf{x} \in P_1 \cup P_2 \cup P_3, \\ 1.4, & \text{if } \mathbf{x} \in P_4 \cup P_5 \cup P_6, \end{cases} \tag{2.64}$$

with $\alpha > 0$. So the equilibria are at $\mathbf{x}^*_{eq_i} = (f_1, f_2, 0)^T$, with $i = 1, \ldots, \eta$, i.e., $\mathbf{x}^*_{eq_1} = (-\alpha, 0, 0)^T \in P_1$, $\mathbf{x}^*_{eq_2} = (\alpha, 0, 0)^T \in P_2$, $\mathbf{x}^*_{eq_3} = (3\alpha, 0, 0)^T \in P_3$, $\mathbf{x}^*_{eq_4} = (-\alpha, 1.4, 0)^T \in P_4$, $\mathbf{x}^*_{eq_5} = (\alpha, 1.4, 0)^T \in P_5$, and $\mathbf{x}^*_{eq_6} = (3\alpha, 1.4, 0)^T \in P_6$. The stable and the unstable manifolds are given by

$$W^s_{\mathbf{x}^*_{eq_i}} = \left\{ \mathbf{x} + \mathbf{x}^*_{\mathbf{eqi}} \in P_i \subset \mathbb{R}^3 \,|\, \mathbf{x} \in span\{V_1\} \right\},$$
$$W^u_{\mathbf{x}^*_{eq_i}} = \left\{ \mathbf{x} + \mathbf{x}^*_{\mathbf{eqi}} \in P_i \subset \mathbb{R}^3 \,|\, \mathbf{x} \in span\{V_2, V_3\} \right\},$$

where $i = 1, \ldots, 6$. The set of eigenvectors is given as follows

$$\left\{ V_1 = (1, 0, 1/2)^T, V_2 = (0, -1, 0)^T, V_3 = (-1, 0, 1)^T \right\}.$$

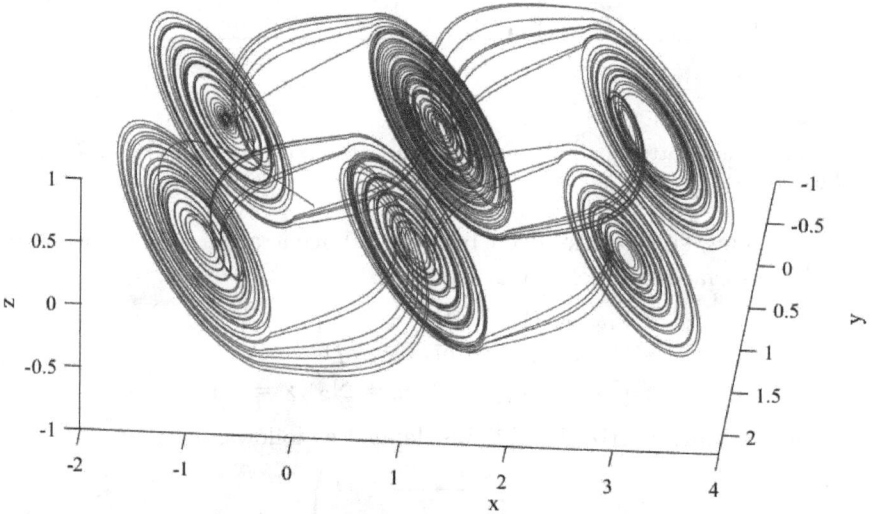

Fig. 2.26   2D-grid multi-scroll attractor of the dynamical system given by (2.59), (2.18), (2.62), (2.63) and (2.64) for the parameters $a = 0.2$, $b = 5$, $c = -3$, $\alpha = 1$, and $\mathbf{x}_0 = 0$.

Figure 2.26 shows a two directional grid multi-scroll attractor of the dynamical system given by (2.59), (2.18), (2.62), (2.63) and (2.64) for the parameters $a = 0.2$, $b = 5$, $c = -3$, $\alpha = 1$, and $\mathbf{x}_0 = 0$.

Now, we discuss the generation of 3D-grid multi-scroll attractor by considering 12 atoms of a $\mathcal{P}$ partition containing a saddle equilibrium point in each of them. The 12 atoms are defined for $\mathbf{N_1} = (2, 0, 1)$, $\mathbf{N_2} = (0, 1, 0)$, $\mathbf{N_3} = (0, 0, 1)$, $D_1 = 0$, $D_2 = 4$, $D_3 = 0.7$ and $D_4 = 0.8$ as follows

$$
\begin{aligned}
&\text{if } \mathbf{N_3} \cdot \mathbf{x} \le -D_4, \text{then} \\
P_1 &= \{\mathbf{x} \in \mathbb{R}^3 : x_3 > 0, \ \mathbf{N_1} \cdot \mathbf{x} \le -D_1, \ \mathbf{N_2} \cdot \mathbf{x} \le -D_3\} \\
&\ \cup \{\mathbf{x} \in \mathbb{R}^3 : x_3 \le 0, \ \mathbf{N_1} \cdot \mathbf{x} < -D_1, \ \mathbf{N_2} \cdot \mathbf{x} \le -D_3\}, \\
P_2 &= \{\mathbf{x} \in \mathbb{R}^3 : x_3 > 0, \ \mathbf{N_1} \cdot \mathbf{x} > -D_1, \ \mathbf{N_1} \cdot \mathbf{x} \le -D_2, \ \mathbf{N_2} \cdot \mathbf{x} \le -D_3\} \\
&\ \cup \{\mathbf{x} \in \mathbb{R}^3 : x_3 \le 0, \ \mathbf{N_1} \cdot \mathbf{x} \ge -D_1, \ \mathbf{N_1} \cdot \mathbf{x} < -D_2, \ \mathbf{N_2} \cdot \mathbf{x} \le -D_3\} \\
P_3 &= \{\mathbf{x} \in \mathbb{R}^3 : x_3 > 0, \ \mathbf{N_1} \cdot \mathbf{x} > -D_2, \ \mathbf{N_2} \cdot \mathbf{x} \le -D_3\} \\
&\ \cup \{\mathbf{x} \in \mathbb{R}^3 : x_3 \le 0, \ \mathbf{N_1} \cdot \mathbf{x} \ge -D_2, \ \mathbf{N_2} \cdot \mathbf{x} \le -D_3\}, \\
P_4 &= \{\mathbf{x} \in \mathbb{R}^3 : x_3 > 0, \ \mathbf{N_1} \cdot \mathbf{x} \le -D_1, \ \mathbf{N_2} \cdot \mathbf{x} > -D_3\} \\
&\ \cup \{\mathbf{x} \in \mathbb{R}^3 : x_3 \le 0, \ \mathbf{N_1} \cdot \mathbf{x} < -D_1, \ \mathbf{N_2} \cdot \mathbf{x} > -D_3\}, \\
P_5 &= \{\mathbf{x} \in \mathbb{R}^3 : x_3 > 0, \ \mathbf{N_1} \cdot \mathbf{x} > -D_1, \ \mathbf{N_1} \cdot \mathbf{x} \le -D_2, \ \mathbf{N_2} \cdot \mathbf{x} > -D_3\} \\
&\ \cup \{\mathbf{x} \in \mathbb{R}^3 : x_3 \le 0, \ \mathbf{N_1} \cdot \mathbf{x} \ge -D_1, \ \mathbf{N_1} \cdot \mathbf{x} < -D_2, \ \mathbf{N_2} \cdot \mathbf{x} > -D_3\} \\
P_6 &= \{\mathbf{x} \in \mathbb{R}^3 : x_3 > 0, \ \mathbf{N_1} \cdot \mathbf{x} > -D_2, \ \mathbf{N_2} \cdot \mathbf{x} > -D_3\} \\
&\ \cup \{\mathbf{x} \in \mathbb{R}^3 : x_3 \le 0, \ \mathbf{N_1} \cdot \mathbf{x} \ge -D_2, \ \mathbf{N_2} \cdot \mathbf{x} > -D_3\}. \\
&\text{if } \mathbf{N_3} \cdot \mathbf{x} > -D_4, \text{then} \\
P_7 &= \{\mathbf{x} \in \mathbb{R}^3 : x_3 > 0, \ \mathbf{N_1} \cdot \mathbf{x} \le -D_1, \ \mathbf{N_2} \cdot \mathbf{x} \le -D_3\} \\
&\ \cup \{\mathbf{x} \in \mathbb{R}^3 : x_3 \le 0, \ \mathbf{N_1} \cdot \mathbf{x} < -D_1, \ \mathbf{N_2} \cdot \mathbf{x} \le -D_3\}, \\
P_8 &= \{\mathbf{x} \in \mathbb{R}^3 : x_3 > 0, \ \mathbf{N_1} \cdot \mathbf{x} > -D_1, \ \mathbf{N_1} \cdot \mathbf{x} \le -D_2, \ \mathbf{N_2} \cdot \mathbf{x} \le -D_3\} \\
&\ \cup \{\mathbf{x} \in \mathbb{R}^3 : x_3 \le 0, \ \mathbf{N_1} \cdot \mathbf{x} \ge -D_1, \ \mathbf{N_1} \cdot \mathbf{x} < -D_2, \ \mathbf{N_2} \cdot \mathbf{x} \le -D_3\} \\
P_9 &= \{\mathbf{x} \in \mathbb{R}^3 : x_3 > 0, \ \mathbf{N_1} \cdot \mathbf{x} > -D_2, \ \mathbf{N_2} \cdot \mathbf{x} \le -D_3\} \\
&\ \cup \{\mathbf{x} \in \mathbb{R}^3 : x_3 \le 0, \ \mathbf{N_1} \cdot \mathbf{x} \ge -D_2, \ \mathbf{N_2} \cdot \mathbf{x} \le -D_3\}, \\
P_{10} &= \{\mathbf{x} \in \mathbb{R}^3 : x_3 > 0, \ \mathbf{N_1} \cdot \mathbf{x} \le -D_1, \ \mathbf{N_2} \cdot \mathbf{x} > -D_3\} \\
&\ \cup \{\mathbf{x} \in \mathbb{R}^3 : x_3 \le 0, \ \mathbf{N_1} \cdot \mathbf{x} < -D_1, \ \mathbf{N_2} \cdot \mathbf{x} > -D_3\}, \\
P_{11} &= \{\mathbf{x} \in \mathbb{R}^3 : x_3 > 0, \ \mathbf{N_1} \cdot \mathbf{x} > -D_1, \ \mathbf{N_1} \cdot \mathbf{x} \le -D_2, \ \mathbf{N_2} \cdot \mathbf{x} > -D_3\} \\
&\ \cup \{\mathbf{x} \in \mathbb{R}^3 : x_3 \le 0, \ \mathbf{N_1} \cdot \mathbf{x} \ge -D_1, \ \mathbf{N_1} \cdot \mathbf{x} < -D_2, \ \mathbf{N_2} \cdot \mathbf{x} > -D_3\} \\
P_{12} &= \{\mathbf{x} \in \mathbb{R}^3 : x_3 > 0, \ \mathbf{N_1} \cdot \mathbf{x} > -D_2, \ \mathbf{N_2} \cdot \mathbf{x} > -D_3\} \\
&\ \cup \{\mathbf{x} \in \mathbb{R}^3 : x_3 \le 0, \ \mathbf{N_1} \cdot \mathbf{x} \ge -D_2, \ \mathbf{N_2} \cdot \mathbf{x} > -D_3\}.
\end{aligned}
$$

$$(2.65)$$

The switching surface are given between adjacent atoms, therefore four switching surfaces are defined, three as previously were defined for 2D grid scroll attractor, and one switching surface between the six first atoms $P_1, \ldots, P_6$, and the six last atoms $P_7, \ldots, P_{12}$. The four switching surfaces are given as follows:

$$
\begin{aligned}
SW_{12} &= SW_{45} = SW_{78} = SW_{10,11} = \mathbf{N_1} \cdot \mathbf{x} = D_1, \\
SW_{23} &= SW_{56} = SW_{89} = SW_{11,12} = \mathbf{N_1} \cdot \mathbf{x} = D_2, \\
SW_{14} &= SW_{25} = SW_{36} = SW_{7,10} = SW_{8,11} = SW_{9,12} = \mathbf{N_2} \cdot \mathbf{x} = D_3, \\
SW_{17} &= SW_{28} = SW_{39} = SW_{4,10} = SW_{5,11} = SW_{6,12} = \mathbf{N_3} \cdot \mathbf{x} = D_4.
\end{aligned}
\tag{2.66}
$$

The constant matrix $B \in \mathbb{R}^{3\times3}$ is defined equally to matrix $-A$ as follows:

$$
B = \begin{pmatrix}
-\dfrac{(a+2c)}{3} & -b & -\dfrac{2(c-a)}{3} \\
\dfrac{b}{3} & -a & -\dfrac{2b}{3} \\
\dfrac{a-c}{3} & b & -\dfrac{(2a+c)}{3}
\end{pmatrix} = -A,
\tag{2.67}
$$

and $\mathbf{F}(\mathbf{x}) = (f_1(\mathbf{x}), f_2(\mathbf{x}), f_3(\mathbf{x}))^T$, where the functional f_1 with $\alpha = 1$ is given by

$$
f_1(\mathbf{x}) = \begin{cases}
-1, & \text{if } \mathbf{x} \in P_1 \cup P_4 \cup P_7 \cup P_{10}, \\
1, & \text{if } \mathbf{x} \in P_2 \cup P_5 \cup P_8 \cup P_{11}, \\
3 & \text{if } \mathbf{x} \in P_3 \cup P_6 \cup P_9 \cup P_{12}.
\end{cases}
\tag{2.68}
$$

And the functional f_2 is given by

$$
f_2(\mathbf{x}) = \begin{cases}
0, & \text{if } \mathbf{x} \in P_1 \cup P_2 \cup P_3 \cup P_7 \cup P_8 \cup P_9, \\
1.4, & \text{if } \mathbf{x} \in P_4 \cup P_5 \cup P_6 \cup P_{10} \cup P_{11} \cup P_{12}.
\end{cases}
\tag{2.69}
$$

Lastly, the functional f_3 is given by

$$
f_3(\mathbf{x}) = \begin{cases}
0, & \text{if } \mathbf{x} \in P_1 \cup P_2 \cup P_3 \cup P_4 \cup P_5 \cup P_6, \\
1.6, & \text{if } \mathbf{x} \in P_7 \cup P_8 \cup P_9 \cup P_{10} \cup P_{11} \cup P_{12}.
\end{cases}
\tag{2.70}
$$

Therefore, the equilibria are at $\mathbf{x}^*_{eq_i} = (f_1, f_2, f_3)^T$, with $i = 1, \ldots, 12$, i.e., $\mathbf{x}^*_{eq_1} = (-1,0,0)^T \in P_1$, $\mathbf{x}^*_{eq_2} = (1,0,0)^T \in P_2$, $\mathbf{x}^*_{eq_3} = (3,0,0)^T \in P_3$, $\mathbf{x}^*_{eq_4} = (-1,1.4,0)^T \in P_4$, $\mathbf{x}^*_{eq_5} = (1,1.4,0)^T \in P_5$, $\mathbf{x}^*_{eq_6} = (,1.4,0)^T \in P_6$. $\mathbf{x}^*_{eq_7} = (-1,0,1.6)^T \in P_7$, $\mathbf{x}^*_{eq_8} = (1,0,1.6)^T \in P_8$, $\mathbf{x}^*_{eq_9} = (3,0,1.6)^T \in P_9$, $\mathbf{x}^*_{eq_{10}} = (-1,1.4,1.6)^T \in P_{10}$, $\mathbf{x}^*_{eq_{11}} = (1,1.4,1.6)^T \in P_{11}$, $\mathbf{x}^*_{eq_{12}} = (3,1.4,1.6)^T \in P_{12}$.

The stable and the unstable manifolds are given by

$$
\begin{aligned}
W^s_{\mathbf{x}^*_{eq_i}} &= \left\{ \mathbf{x} + \mathbf{x}^*_{\mathbf{eq_i}} \in P_i \subset \mathbb{R}^3 \,|\, \mathbf{x} \in span\{V_1\} \right\}, \\
W^u_{\mathbf{x}^*_{eq_i}} &= \left\{ \mathbf{x} + \mathbf{x}^*_{\mathbf{eq_i}} \in P_i \subset \mathbb{R}^3 \,|\, \mathbf{x} \in span\{V_2, V_3\} \right\},
\end{aligned}
$$

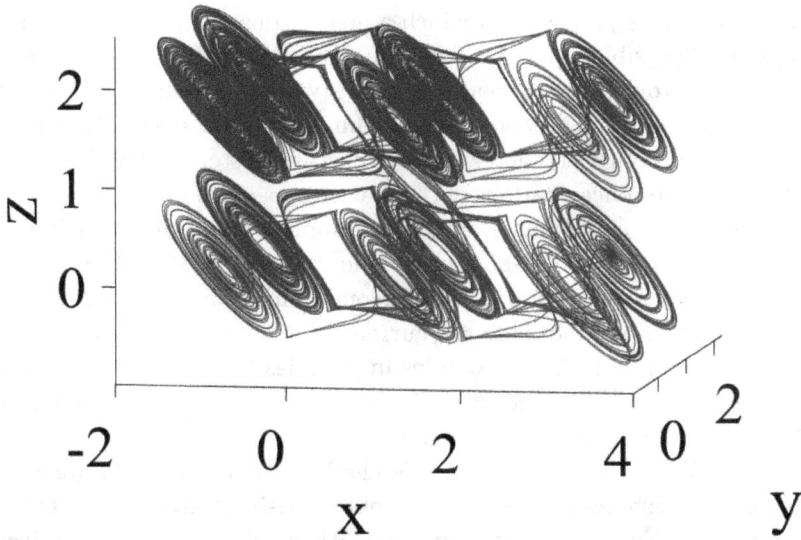

Fig. 2.27   3D-grid multi-scroll attractor of the dynamical system given by (2.59), (2.18), (2.67), (2.68), (2.69) and (2.70) for the parameters $a = 0.2$, $b = 5$, $c = -3$, $\alpha = 1$, and initial condition $\mathbf{x}_0 = 0$.

where $i = 1, \ldots, 12$. The set of eigenvectors is given as follows

$$\left\{ V_1 = (1, 0, 1/2)^T, V_2 = (0, -1, 0)^T, V_3 = (-1, 0, 1)^T \right\}.$$

Fig. 2.27 shows a three-dimensional grid multi-scroll attractor of the dynamical system given by (2.59), (2.18), (2.67), (2.68), (2.69) and (2.70) for the parameters $a = 0.2$, $b = 5$, $c = -3$, $\alpha = 1$, and initial condition $\mathbf{x}_0 = 0$.

2.6   Multi-stability in PWL systems

In the evolution of a complex system, there are several possible (coexisting) basins of attraction with a sink that traps the system trajectory depending on its initial state. This phenomenon is usually called multi-stability and appears in a wide variety complex systems [Campos-Cantón *et al.* (2010); Angeli (2007)]. The correct interpretation of a sink depends on the complex system being studied. For instance, in the context of biology there are many examples of systems that manifest multi-stability phenomena. An example worth mentioning is the cellular differentiation in order to understand human development and distinct forms of diseases. Here, multi-stability is understood as a processes in which a gene regulation network alternates along several possible types of cell [Ghaffarizadeh *et al.* (2014)]. Another example comes from the nonlinear chemical dynamics, where multi-stability is understood as the different possible final chemical states [Sagués and Epstein (2003)]. In this context, the archetype system is the Oregonator oscillator, where concentrations of the reacting species oscillate between two stable final states (bistability). Several examples can be cited ranging from medicine [Haddad *et al.* (2011)], electronic [Patel *et al.* (2014)], visual perception [Gershman *et al.* (2012)], superconducting [Jung *et al.* (2014)], etc. All of these examples motivate the current research works that address the challenger posed by R. Vilela Mendes in [Mendes (2000)] of identifying the universal mechanism that leads to multi-stability and to prove rigorously under what circumstances the phenomenon may occur.

One feasible mode to address this challenge is through the formalism of dynamical systems where the concepts of basin of attraction, stability, convergence among others have a mathematical definition and also allow us to use some tools from stability theory to analyze its behavior. It is worth to note that this situation is similar with the research works some year ago where chaotic behavior was modeled and interpreted from the point of view of dynamical systems. Since then, various dynamical systems with a chaotic behavior have been proposed (some examples are the Lorenz, Chua and Rössler systems, to name a few).

In the context of dynamical systems, an attractor is defined as a subset of the phase space toward which the trajectories of the dynamical systems converge to it. Attractors can be fixed points, limit cycles, quasiperiodic, chaotic or hyper-chaotic orbits. The basin of attraction is defined as the set of all the initial conditions in the phase space whose corresponding trajectories converge to an attractor [Kengne (2017); Giesl (2007)]. Concepts of convergent trajectories and attractor stability are usually associated with a energy-like term called Lyapunov function. Then, with the above concepts it can be said that a multi-stable dynamical system is a dynamical system that, depending on its initial condition, its trajectories can converge to two or more mutually exclusive Lyapunov stable attractors [Haddad *et al.* (2011)].

Some formal definitions of multi-stable behavior have been proposed by D. Angeli in [Angeli (2007)] and Q. Hui in [Hui (2014)] for discontinuous dynamical systems. On the other hand, some methodologies to induce a multi-stable behavior by coupling two or more dynamical systems have been reported. For example, E. Jiménez-López *et al.* have generated multistable behavior by employing a pair of Unstable Dissipative Systems (UDS) of Type I, coupled in a master-slave configuration [Jiménez-López *et al.* (2013)]. It is worth to mention that an UDS is a Piecewise Linear System (PWL) which is classified in two types according to the eigenvalues of the linear operator. On the other hand, C. R. Hens *et al.* have shown that two coupled Rössler oscillators can achieve a certain type of multi-stability called extreme, where the number of coexisting attractors is infinite [C. R. Hens, R. Banerjee, U. Feudel and Dana (2012)]. It has also been observed that by an appropriate modification of the equations, some classical chaotic systems can exhibit also a multistable behavior. For example, in [Kengne (2017); Kengne *et al.* (2016)] J. Kengne *et al.* have proposed a system based on the Duffing-Holmes system and Chua's oscillator. They have shown that in a given range of its parameter values this system exhibit coexisting attractors. Additionally, H. E. Gilardi-Velázquez *et al.* introduced a multi-stable system generated with a Piecewise Linear (PWL) system based on UDS type I and the jerk equation, in which the switching among the different phase space regions is driven by means of the Nearest Integer or the round(x) function, such that the system display infinite attractors along one dimension, the aim of its approach is to drive the linear operator close to be stable. [Gilardi-Velázquez *et al.* (2016)]. The experimental evidence of multi-stability for the Rössler oscillator have been reported by M. Patel *et al.* in [Patel *et al.* (2014)]. On the other hand, C. Li in [Li and Sprott (2013)] and D. Z. T. Njitacke *et al.* in [Z. T. Njitacke, J. kengne, H. B. Fotsin, A. Nguomkam Negou (2016)], have been observed that multi-stable behavior is also presented in the Butterfly-Flow system and in the memristive diode bridge-based Jerk circuit, respectively. It is worth to mention that for some discrete-time chaotic systems, the multi-stable behavior is also displayed [Carvalho *et al.* (2001); Astakhov V, Shabunin A, Uhm W (2001)].

Let $\phi_t(\chi_0) \in \mathbf{R}^n$ be the solution curve or trajectory of (2.1) given the initial condition χ_0. Thus $\phi : \mathbb{R}^n \times \mathbb{R} \to \mathbb{R}^n$:

Definition 2.7. A closed invariant set $\mathcal{A} \subseteq \mathbb{R}^n$ is called an attracting set of (2.1) with flow ϕ_t, if there exist a neighborhood $U \subseteq \mathbb{R}^n$ of $\mathcal{A}$ with $\phi_t(U) \subseteq U$ and $\mathcal{A} \subseteq U$ such that

$$\mathcal{A} = \bigcap_{t=0}^{\infty} \phi_t(U),$$

where

$$\phi_t(U) = \{\phi_t(x) | x \in U\}.$$

An attractor of (2.1) is an attracting set which contains a dense orbit.

Definition 2.8. The basin of attraction of $\mathcal{A}$ is the set of initial conditions whose trajectories converge to the attractor, that is $U = \Omega(\mathcal{A}) = \{\chi_0 \in \mathbb{R}^n : \phi_t(\chi_0) \to \mathcal{A}$ as $t \to \infty\}$.

Thus, an attractor is a closed invariant set $\mathcal{A}$ and there is an open neighborhood $U \supset \mathcal{A}$ such that the trajectory $\chi(t) = \phi_t(\chi_0)$ of any point $\chi_0 \in U$ satisfies $d(\phi_t(\chi_0), \mathcal{A}) \to 0$ as $t \to \infty$; where $d(\chi, \mathcal{A}) = \inf\{d(\chi, x_0)| \, \chi \in \phi_t(\chi_0)$ and $x_0 \in \mathcal{A}\}$.

There are different types of attractors, i.e., stable equilibrium point, limit cycle, a set generated by a chaotic trajectory.

Definition 2.9. We say that the dynamical system given by (2.1) is generalized multistable if there exist more than one basin of attraction, i.e., $\Omega(\mathcal{A}_1), \ldots, \Omega(\mathcal{A}_k)$, with $2 \leq k \in \mathbb{Z}$.

In the context of generalized multi-stability, the coexistence of multiple attractors $\mathcal{A}_1, \ldots, \mathcal{A}_k$ makes the distance $d(\phi_t(\chi_0), \mathcal{A}_i)$ takes different values that depends on the initial condition χ_0. For instance, the distance $d(\phi_t(\chi_0), \mathcal{A}_i) = 0$ if $\chi_0 \in \Omega(\mathcal{A}_i)$, but $d(\phi_t(\chi_0), \mathcal{A}_i) \neq 0$ if $\chi_0 \in \Omega(\mathcal{A}_j)$, with $i \neq j$.

Remark 2.2. It is important to characterize different types of multistability as follows:

1. There exists a set $\{\chi_i^*\}_{i=1}^m$ of saddle equilibrium points of (2.1) in $\mathbf{R}^n$. The basin of attraction of each equilibria is given by the stables set E^s. This type of multi-stable states is known as multi-stability.

2. Due to the phase space $\mathbb{R}^n$ is partitioned in a finite number $m \in \mathbb{Z}$ of domains S_i and each equilibrium point is located at $\chi_i^* = A^{-1}B_i \in S_i \subseteq \mathbb{R}^n$, for $i = 1, \ldots, m$. So the basins of attraction of χ_i^* is determined by the stable set E_i^s restricted to S_i.

3. When the trajectory does not converge to the equilibria, instead oscillates around them and exist at least two basin of attractors $\Omega_i = \Omega(\mathcal{A}_i)$ and $\Omega_j = \Omega(\mathcal{A}_j)$, this type of multi-stable states is known as generalized multi-stability. This is the target of this work.

2.6.1   *Emerging multi-stability in a multi-scroll attractor based on UDS Type I*

Now based on the previous method for generating multi-scroll attractors in Section 2.3.1, the following linear operator and constant vector are considered:

$$\mathbf{A} = \begin{pmatrix} 0 & 1 & 0 \\ 0 & 0 & 1 \\ -10.5 * \nu & -7.0 * \nu & -0.7 * \nu \end{pmatrix}, \quad \mathbf{B} = \begin{pmatrix} 0 \\ 0 \\ \nu * b_3 \end{pmatrix}, \quad (2.71)$$

where $\nu \in \mathbb{R}^+$ is a constant parameter.

With this vector $\mathbf{B}$ the displacement of the equilibria is also along the x_1 axis, where b_3 switches according to the *round* function given by Eq. (2.37).

The parameter ν is used to change the eigenspectra of the linear part of systems, mainly the directions of the stable and unstable manifolds, so ν can be taken as a bifurcation parameter. Figures 2.28(a) and (b) show the bifurcation diagram by means of considering the local maxima at every domain $\mathcal{D}_i$ of each scroll depicted at x_1 (i.e. if the flow $\phi_{x_1}^{t_j} \geq \phi_{x_1}^{t_n} \in \mathcal{D}_i$ with $t_j > t_n$) for the range of the parameter $0 \leq \nu \leq 2$ and a zoomed area at $0.95 \leq \nu \leq 1.5$. Both diagrams were calculated by the same initial condition $\mathbf{X}_0 = (0.7, 0, 0)^T$, the difference is that Fig. 2.28(a) was calculated for 10,000 iterations while Fig. 2.28(b) for 1,000,000 iterations. It is important to note that for $\nu < 1.1$ the local maxima increases according with time. Otherwise, for $\nu \geq 1.1$ the local maxima remain constant regardless of time. This shows the dynamic transition between escape time to confinement. Additionally Figs. 2.28(c) and (d) show the number of domains $\mathcal{D}_i$ that the system visited for the same values of ν, respectively. Notice that, if the range of $\nu \geq 1.1$ approximately, the number of domains $\mathcal{D}_i$ visited remain at the constant value of 3, because the system presents only a single-scroll attractor located in the inside domain as depicted in Fig. 2.29. This can be better appreciated in the zoomed area in Fig. 2.28(d).

With these parameter values the equilibria of the system is given by $\mathbf{X}_0^* = (b_3/10.5, 0, 0)^T$. Therefore $c = 6.3$ and $\alpha = c/10.5$ are assigned to the function in Eq. (2.37). The eigenspectrum of the matrix $\mathbf{A}$ depends now on the bifurcation parameter ν.

Now, in order to understand the phenomenon and visualize the exact location of the intersection of E^s and E^u along with the points belonging to the attractor $\mathfrak{A}$ at the commutation surface $x_1 = x_{1_{cs}}$, a Poincaré plane was implemented exactly at the commutation surface. First, the Poincaré plane is defined as $\Sigma := \{(x_1, x_2, x_3) \in \mathbb{R}^3 : \mu_1 x_1 + \mu_2 x_2 + \mu_3 x_3 + \mu_4 = 0\}$, where $\mu_1, \ldots, \mu_4 \in \mathbb{R}$ are the coefficients of a plane equation whose values are considered depending on the location under study, which in this case it will be in the commutation surfaces $x_1 = x_{i_{cs}}$ with $i \in \mathbb{Z}$ guaranteeing $\mathfrak{A} \cap \Sigma \neq \emptyset$. The crossing events of interest

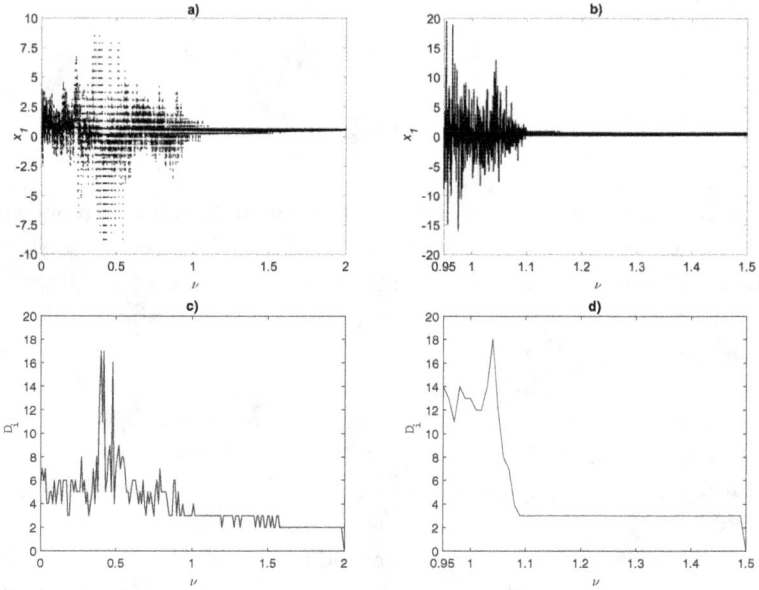

Fig. 2.28   (a) Bifurcation diagram of the system given by Eqs. (2.1) with (2.37) and (2.71) for the value of $0 \leq \nu \leq 2$ for Fig. (b) depicts the bifurcation for the range $0.95 \leq \nu \leq 1.5$ for 1,000,000 iterations. Figures (c) and (d) show the number of domains $\mathcal{D}_i$ visited by the trajectory of the systems for the same values of the bifurcation parameters above. The initial condition considered for both diagrams is $\mathbf{X_0} = (0.7, 0, 0)^T$.

are $\{\phi_{in}^{t_1}(\mathbf{X_0}), \phi_{out}^{t_2}(\mathbf{X_0}), \phi_{in}^{t_3}(\mathbf{X_0}), \phi_{out}^{t_4}(\mathbf{X_0}), \ldots, \phi_{in}^{t_{m-1}}(\mathbf{X_0}), \phi_{out}^{t_m}(\mathbf{X_0})\} \in \Sigma$ with $m \in \mathbb{Z}^+$. Where m corresponds to the total of crossing events in Σ, and ϕ_f^j correspond to the j-th intersection of $\mathfrak{A} \cap \Sigma$ in the $f = in, out$ direction. The sub-index *out* corresponds to trajectories that cross Σ with $dx_1/dt > 0$, and *in* corresponds to trajectories that cross Σ with $dx_1/dt < 0$.

Figures 2.30(a) and (b) present the plane of $\Sigma := \{(x_1, x_2, x_3) \in \mathbb{R}^3 : x_1 - x_{1_{cs}} = 0\}$ for the values of $\nu = 1$ and $\nu = 1.42$, respectively. Three issues can be noticed from these intersections.

I. Distribution of the intersections with the Poincaré plane

Consider a trajectory in $\mathfrak{A} \cap D_i$, with equilibrium point $X_i^* \in D_i$. When the trajectory $\phi^t(\mathbf{X_0})$ oscillates around X_i^* increases the distance to the equilibrium point and exits the current domain $\mathcal{D}_i$ to $\mathcal{D}_{i+1}$ (or $\mathcal{D}_i$ to $\mathcal{D}_{i-1}$) through the commutation surface near the region of intersection of the

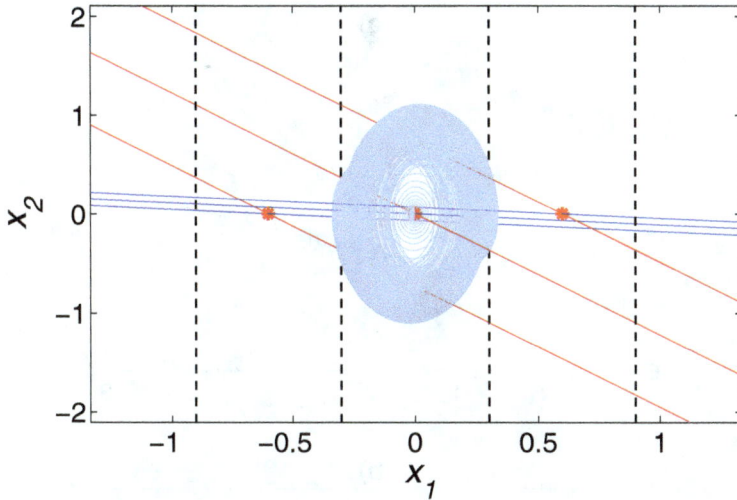

Fig. 2.29   Projection of the trajectory of the system (2.71) with (2.37) onto the (x_1, x_2) plane with $c = 6.3$, $\alpha = 0.6$ and $\nu = 1.42$. Marked with red asterisk the equilibria of the system, and with gray line the commutation surfaces generated by the function (2.37). The initial condition of the system is $\mathbf{X}_0 = (-0.1, 0, 0)^T$.

unstable manifold with the Poincaré plane $(E^u \cap \Sigma)$, as it can be appreciated with the intersection points of the trajectory $\phi_{out}^{t_j}(\mathbf{X}_0)$ marked in blue circles. The blue triangle corresponds to the intersection of the vector $Real(\vartheta_2(\nu))$ with the Poincaré plane, i.e., $Real(\vartheta_2(\nu)) \cap \Sigma$. The blue line appearing in both Figs. 2.30(a) and (b), corresponds to the intersection of the unstable manifold with the Poincaré plane $E^u \cap \Sigma$. Notice that most of the trajectories are crossing near this section due to the scrolling behavior in or near the unstable manifold, i.e., the majority of the blue circles are located along the intersection marked with the blue line. Nevertheless, Fig. 2.30(a) presents a region of escaping intersection points $\phi_{out}^{t_j}(\mathbf{X}_0)$ not near the unstable manifold (blue circles not located in the intersection marked in blue line), which comes out as one of the main difference between the escaping points in Fig. 2.30(b) for a larger value of the bifurcation parameter. Also notice that the entering intersection points $\phi_{in}^{t_j}(\mathbf{X}_0)$, marked with orange asterisks for $\nu = 1$ are scattered around the intersection of the stable manifold and the Poincaré plane $E^s \cap \Sigma$ marked with the red triangle in Fig. 2.30(a), but in Fig. 2.30(b), the crossing events are located below this intersection in an apparently ranked or aligned way. Considering this change in the crossing events, it is easy to determine that the

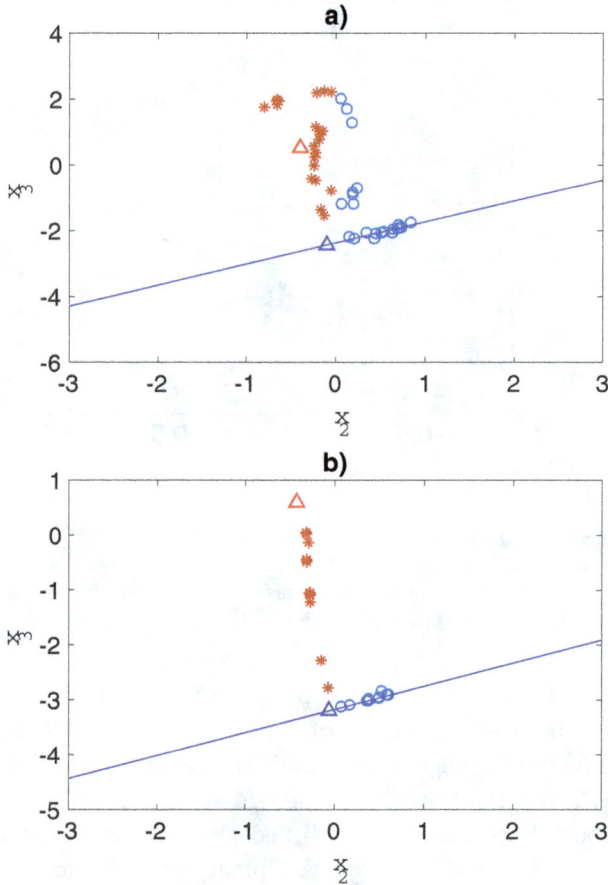

Fig. 2.30   Intersections of the trajectory of the system (2.1) with (2.37) and (2.71) with the commutation surface Σ located at $x_{1_{cs}}$ for (a) $\nu = 1$, (b) $\nu = 1.42$. Marked with blue circles the trajectories exiting $\mathcal{D}_1$ near $E^u \cap \Sigma$. The orange asterisks represent the trajectories entering $\mathcal{D}_1$. The red triangle stands for $E^s \cap \Sigma$, and the blue line corresponds to the intersection of unstable manifold and commutation surface.

distance between the trajectory crossing events and the intersection points of the manifolds is decreasing (Note the ranges of values taken by the x_3 axis in Figs. 2.30(a) and (b)). To estimate this approach the following considerations were made.

Consider the system given by Eq. (2.71), with $c = 6.3$ and $\alpha = 0.6$. The spectrum of the matrix $\mathbf{A}$ depends on the bifurcation parameter ν, which for the value of $\nu = 1.42$ is given by:

$$\lambda(\nu) = \{-1.4164, 0.2082 \pm i3.2475\},$$

$$\vartheta(\nu) = \{\vartheta_{1,2,3}(\nu)\} = \left\{ \begin{pmatrix} 0.3771 \\ -0.5342 \\ 0.7566 \end{pmatrix} \begin{pmatrix} 0.0892 \pm i0.0115 \\ 0.0187 \pm i0.2919 \\ -0.9520 \end{pmatrix} \right\}, \qquad (2.72)$$

proving that the Definition 2.5 is satisfied. Considering these values, the system results in an interesting multistable state phenomena due to the round function and the location of their eigenvectors as depicted in Fig. 2.29 for the initial condition $\mathbf{X}_0 = (-0.1, 0, 0)^T$. The attractor is located near the equilibria in the origin marked in red asterisk due to the initial condition given. However there is no oscillation near the adjacent equilibrium points, the reason of this resides on the eigenvectors, as they are not located in the same way as in the previous examples, i.e., the stable manifold of the domain in witch the trajectory oscillates, doesn't match and cross with the stable manifolds of the adjacent domains.

This can be easily observed in Fig. 2.31 as it is next explained. Figures 2.31(a) and (b) present a projection of a trajectory onto the plane (x_1, x_2) for the systems (2.1) with (2.37) and (2.71), both initialized with the same initial condition near the origin $\mathbf{X}_0 = (-0.1, 0, 0)^T$, but considering the values in the bifurcation parameter of $\nu = 1$ and $\nu = 1.42$, respectively. Notice in the projections that as the time increases the oscillating clockwise trajectory on both systems attractor grows larger, until eventually the trajectories on the scrolls cross the commutation surface close to the unstable manifold of E^u marked in blue lines. Apparently both systems trajectory cross the commutation surface plane $x_1 = x_{1_{cs}}$ marked with black lines in a neighborhood near to the intersection of the unstable and stable manifolds marked with blue and red lines (depending if the trajectory is escaping or entering the domain), respectively. However, by observing different projections of the attractors, for example the projection onto the plane (x_1, x_3) in Figs. 2.31(c) and (d), it can be appreciated that the direction of the manifold has been slightly changed due to the variation of the parameter from $\nu = 1$ to $\nu = 1.42$. Take a closer look at the graph in Fig. 2.31(c) between $-1 < x_1 < -0.3$. In this domain $\mathcal{D}_{-1}$ the trajectory of the system is entering from the upper part close to $x_3 \approx 2$ where the black arrow depicts the direction of the crossing in the intersection of the unstable manifold E^u with the commutation surface $x_1 = x_{-1_{cs}}$. After entering to $\mathcal{D}_{-1}$ the trajectory is directed below the stable manifold E^s in this domain marked with the red line, and then crosses to the domain $\mathcal{D}_0$

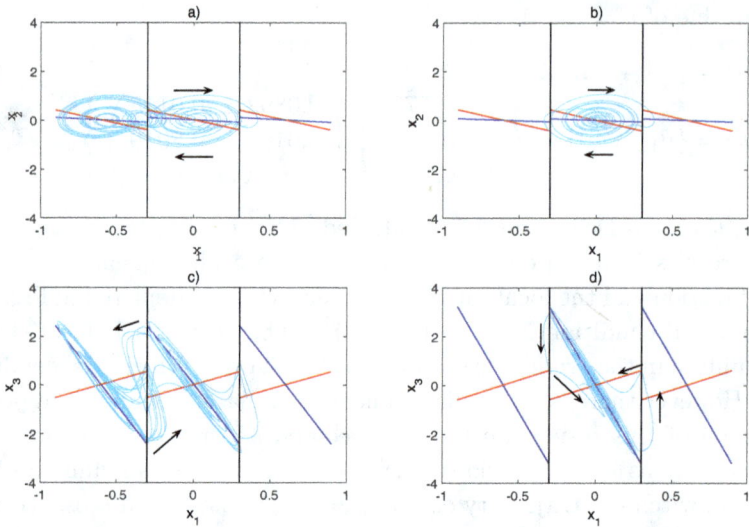

Fig. 2.31   Trajectory projections for the system given by Eq. (2.71), with $c = 6.3$ and $\alpha = 0.6$. For $\nu = 1$ (a) onto the plane (x_1, x_2), (c) onto the plane (x_1, x_3). For $\nu = 1.42$ (b) onto the plane (x_1, x_2), (d) onto the plane (x_1, x_3). Marked with green line the commutation surfaces generated by the function (2.37), with red line the complex eigenvector and with black line the real eigenvector. The black arrows show the trajectory direction.

near the intersection of the next stable manifold in $\mathcal{D}_0$ in $x_3 \approx -0.9$. Notice that some of the trajectories entering this domain end up oscillating into the scroll as the arrow in the lower part depicts (this phenomena can also be seen in the projection of the multi-scroll attractor in Figs. 2.8(b) and (d)). Nevertheless, some of the trajectories instead of reaching the scrolling plane in $\mathcal{D}_0$ are redirected again to $\mathcal{D}_{-1}$ and oscillate in the scroll in the unstable manifold E^u.

This behavior is completely different in the projection of the attractor in Fig. 2.31(d) for $\nu = 1.42$. Here, when the trajectory escapes the domain $\mathcal{D}_0$ near $x_3 \approx 3$ and enters $\mathcal{D}_{-1}$, the trajectory is directed towards the location of the stable manifold and crosses near the intersection $E^s \cap x_{-1_{cs}}$ at approximately $x_3 \approx 0.1$. After crossing the trajectory is redirected towards the scroll in $\mathcal{D}_0$ as the arrows depicts the direction. This process repeats continuously from the three consecutive domains in which the trajectory of the system lies (Notice this also from Fig. 2.28(b)).

This comes as a major advantage in the proposed system as one may consider different initial conditions for values of $1.1 < \nu < 2.1$, and for

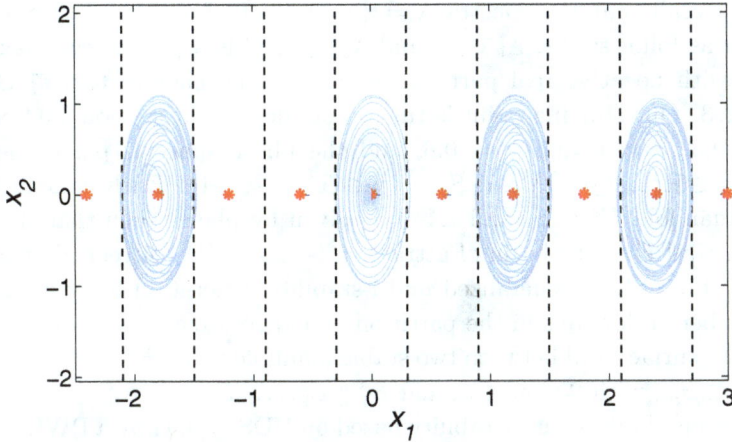

Fig. 2.32 Projections of the trajectories of the system (2.71) with (2.37) onto the (x_1, x_2) plane with $\nu = 1.42$, $c = 6.3$ and $\alpha = 0.6$ for the initial condition sets $\mathbf{X}_0 = (-2.0, 0, 0)^T$, $\mathbf{X}_0 = (0.01, 0, 0)^T$, $\mathbf{X}_0 = (1.0, 0, 0)^T$ and $\mathbf{X}_0 = (2.21, 0, 0)^T$.

each initial location given the system will oscillate near those continuous domains along the x_1 axis from $(-\infty, \infty)$. Consider the different sets of initial conditions $\mathbf{X}_0 = (-2, 0, 0)^T$, $\mathbf{X}_0 = (0.01, 0, 0)^T$, $\mathbf{X}_0 = (1.0, 0, 0)^T$ and $\mathbf{X}_0 = (2.21, 0, 0)^T$ for $\nu = 1.42$. The resulting trajectory of the system given these conditions are depicted in Fig. 2.32, each scroll presented is a unique experiment for a given number of iterations in time. Notice that the orbit does not form attractors on the neighborhood domains. Resulting in a multistable system for the x_1 axis from $(-\infty, \infty)$ which oscillates around the equilibria in which the initial condition is set, and the trajectory is bounded by the dynamics of the left and right equilibrium points although there are no oscillations around them. Additionally, the largest Lyapunov exponent of the attractor was calculated, taking a value of $MLE = 0.113669$ demonstrating a chaotic behavior in the system.

2.6.2   *Emerging multi-stability in a multi-scroll attractor based on UDS Type II*

Now, the interest is to generate multi-stability behavior via a dynamical system based on UDS *Type II*, so we consider the system (2.3) with (2.39).

   The idea of generalized multi-stability generation is different to that given in Section 2.6.1, instead of controlling the stable and unstable manifolds, it is increased the number of domains in the partition of the phase space. Recall that the spectra $\Lambda = \{\lambda_1, \lambda_2, \lambda_3\}$ of the linear operator A is given as follows: $0 < \lambda_1 \in \mathbb{R}$, and $\lambda_2, \lambda_3 \in \mathbb{C}$ is a pair of complex conjugate with negative real part and corresponding eigenvectors $\bar{v}_j \in \mathbb{R}^n$, $j = 1, 2, 3$. Our starting point is the system defined in Section 2.3.2 where $\alpha = -0.6$, $\beta = 6$ and $\gamma = 0.6$, and the phase space is partitioned by $S_1 = \{\chi \in \mathbb{R}^3 | x_1 \geq -1\}$ and $S_2 = \{\chi \in \mathbb{R}^3 | x_1 < -1\}$. Each domain has a stable manifold $E_1^s \subset S_1$ and $E_2^s \subset S_2$ given by planes such that they are parallel $E_1^s \parallel E_2^s$. The basin of attraction is located Ω between E_1^s and E_2^s. So, now the idea of generalized multi-stability generation is by increasing the number of domains in the partition and generate an attractor near the switching surface and between two stable manifolds, i.e., $E_1^s \subset S_1$, $E_2^s \subset S_2$, ..., $E_k^s \subset S_m$, with $2 \leq m \in \mathbb{Z}$, and $E_1^s \parallel E_2^s, \ldots, E_{m-1}^s \parallel E_m^s$.

We exemplify the multistability based on UDS type II by a PWL system which is capable of producing generalized bi-stability. The PWL system given in Section 2.3.2 is used but now the phase space is partitioned in three domains given by $S_{1x_1} = \{\chi \in \mathbb{R}^3 | x_1 > 1\}$, $S_{2x_1} = \{\chi \in \mathbb{R}^3 | -1 \leq x_1 \leq 1\}$ and $S_{3x_1} = \{\chi \in \mathbb{R}^3 | x_1 < -1\}$. The switching surfaces are given by $\Sigma_1 = \{\chi \in \mathbb{R}^3 | x_1 = 1\}$ and $\Sigma_2 = \{\chi \in \mathbb{R}^3 | x_1 = -1\}$. This action of introducing a new domain modifies the function $\sigma(\chi)$ given by (2.40) as follows:

$$\sigma(\chi) = \begin{cases} -7, & \text{if } \chi \in S_{1x_1}; \\ 0, & \text{if } \chi \in S_{2x_1}; \\ 7, & \text{if } \chi \in S_{3x_1}. \end{cases} \tag{2.73}$$

Now, the equilibria are located at $x_1^* = (0, 0, 0)$ and $x_{2,3}^* = (\pm 11.66, 0, 0)$, and this dynamical system presents two attractors $\mathcal{A}_\mathcal{L}$ and $\mathcal{A}_\mathcal{R}$, which are shown in Fig. 2.33. The left-hand side attractor $\mathcal{A}_\mathcal{L}$ and right-hand side attractor $\mathcal{A}_\mathcal{R}$ are generated by considering the following initial conditions: $\chi_{01} = (-1.1, 0, 0)^\top$ and $\chi_{02} = (1.1, 0, 0)^\top$. In terms of generalized multistability we have a bi-stable behavior and the basin of attraction of the system is given by the union of two basins of attraction $\Omega_1 \cup \Omega_2$. Figure 2.34 shows the basins of attraction Ω_1 and Ω_2 corresponding to two attractors $\mathcal{A}_\mathcal{L}$ and $\mathcal{A}_\mathcal{R}$, respectively.

The generalization of bistable to multi-stable behavior given by a dynamical system based on UDS *Type II* can be given by adding more domains to the partition by considering switching surfaces Σ_i perpendicular to the

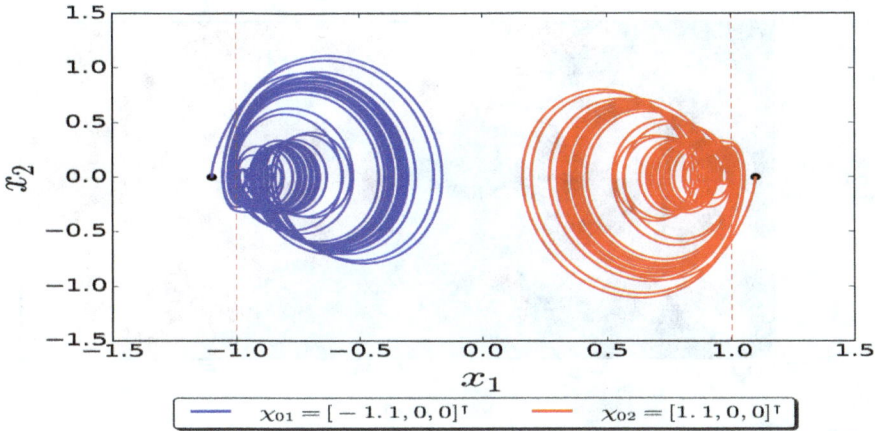

Fig. 2.33 Projections of the attractors based on UDS *Type II* onto the (x_1, x_2) plane, with the switching law (2.73) and $\alpha = 0.6$ and $p = 10$. The black dots indicate initial conditions for the left-hand side attractor $\mathcal{A}_\mathcal{L}$ and right-hand side attractor $\mathcal{A}_\mathcal{R}$, respectively.

axis x_1 based on the aforementioned. But it is not the only way how we can add more domains. The other idea is explained with the next example.

Now we explain how to generate four attractors in a two-dimensional grid (2D-grid scroll attractors) by modifying the piecewise constant vector B. The equilibria of the system is given by $\chi^* = (\sigma(\chi)/\alpha, 0, 0)^\top$ by considering (2.39). Notice that the equilibria are located in the axis x_1 but now we want the equilibria are located onto the plane (x_1, x_2) as follows $\chi^* = (\sigma(\chi)/\alpha, f(\chi), 0)^\top$, thus $B = -A\chi$.

$$B(\chi) = \begin{pmatrix} -f(\chi) \\ 0 \\ \sigma(\chi) + \beta f(\chi) \end{pmatrix},$$ (2.74)

where $\sigma(\cdot)$ is the switching law give by (2.73), and $f(\chi)$ is the following step function:

$$f(\chi) = \begin{cases} -1.4, & \text{if } \chi \in S_{1x_2} = \{\chi \in \mathbf{R}^3 : x_2 \le 0\}; \\ 1.4, & \text{if } \chi \in S_{2x_2} = \{\chi \in \mathbf{R}^3 : x_2 > 0\}. \end{cases}$$ (2.75)

The role of the function $f(\cdot)$ is to split the x_2 direction for each one of the switching surfaces S_{ix_2} and add more domains. Now the space $\mathbf{R}^3$ is partitioned in six domains given as follows: $S_1 = S_{1x_1} \cap S_{1x_2}$, $S_2 = S_{2x_1} \cap S_{1x_2}$,

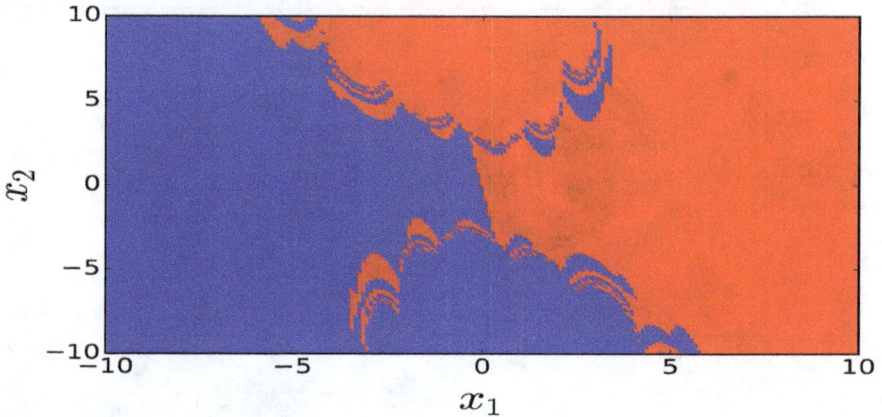

Fig. 2.34   The basins of attraction Ω_i, for $i = 1, 2$, of the system given in Example 2 and the switching function (2.73) in $x_1 \in [-10, 10]$, $x_2 \in [-10, 10]$ and $x_3 = 0$. Blue and red points correspond to initial conditions that converge to the $\mathcal{A}_{\mathcal{L}}$ attractor and the $\mathcal{A}_{\mathcal{R}}$ attractor, respectively.

$S_3 = S_{3x_1} \cap S_{1x_2}$, $S_4 = S_{1x_1} \cap S_{2x_2}$, $S_5 = S_{2x_1} \cap S_{2x_2}$, $S_6 = S_{3x_1} \cap S_{2x_2}$. Now there are six equilibria located at: $x_{1,3,4,5}^* = (\pm 11.66, \pm 2, 0)$ and $x_{2,5}^* = (0, \pm 2, 0)$, with the three equilibrium points added is possible generate four attractors. Figure 2.35 shows multi-stable behavior for the four coexisting attractors, which are generated by using the following initial conditions: $\chi_{01} = (-1.1, 2, 0)^\top$, $\chi_{02} = (1.1, 2, 0)^\top$, $\chi_{03} = (-1.1, -2, 0)^\top$ and $\chi_{04} = (1.1, -2, 0)^\top$. Each final stable state of the system is a single chaotic attractor which depend only of the initial condition selected. In Fig. 2.36, basins of attraction of the system based on UDS *Type II* is shown by varying the initial conditions in the range $x_1 \in [-10, 10]$, $x_2 \in [-10, 10]$ and $x_3 = 0$. As with UDS *Type I* is possible to extend the number of final states by adding more equilbria along x_2 with switching domains.

2.6.3   *Multi-stable attractors generated by the two types of UDS*

In practice, the possibility of converting a multi-stable system to a mono-stable one is very much in demand because this would allows us to avoid any unpredictable switch to another coexisting state that may be caused by environmental fluctuations or increasing internal noise [Sevilla-Escoboza *et al.* (2015)].

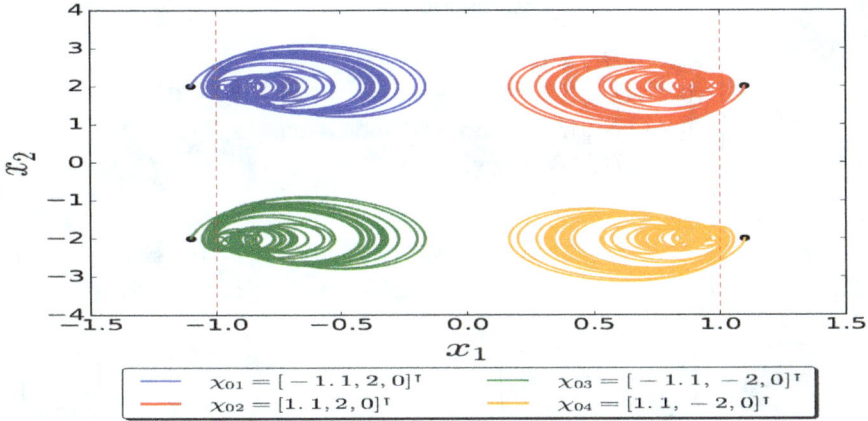

Fig. 2.35   Projections of the attractors based on UDS *Type II* onto the (x_1, x_2) plane with $\alpha = 0.7$; $p = 10$ and, vector-valued function B given by (2.74).

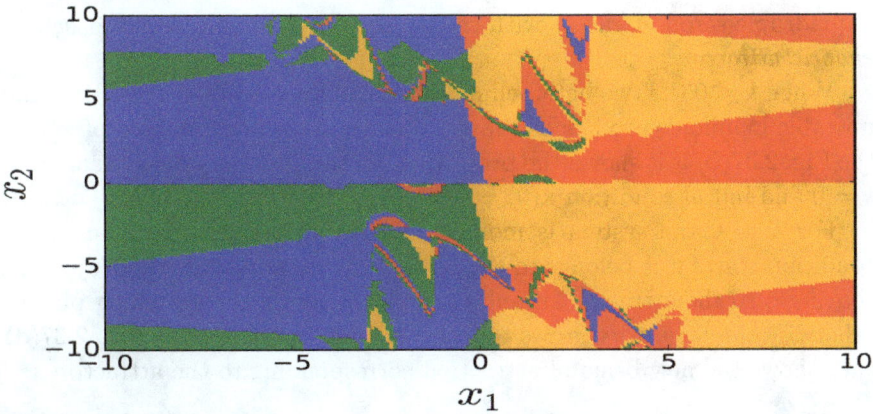

Fig. 2.36   Basins of attraction, of the attractors based on UDS *Type II*, onto the (x_1, x_2) plane, considering $x_1 \in [-10, 10]$, $x_2 \in [-10, 10]$ and $x_3 = 0$. and, vector-valued function B given by (2.74). Each point represents a given initial condition and its color is the basin of attractions in which the UDS converge with such initial condition.

In this section, a parametric control is introduced in the system previously introduced in Section 2.3.4. The proposed control allows us to annihilate the bistable behavior trough the location of the heteroclinic orbits. When the heteroclinic orbits of both attractors $\mathcal{A}_{\mathcal{L}}$ and $\mathcal{A}_{\mathcal{R}}$ are close enough, the attractors fuse themselves in a quadruple-scroll attractor. For a specific parameter value the two heteroclinic orbits collide and merge together.

So the PWL system is given as follows:

$$\dot{\mathbf{x}} = \begin{cases} A_1\mathbf{x} + B_1, & \text{if } x_1 + x_3 < 0 \text{ and } -x_1 + x_3 < 0; \\ A_2\mathbf{x} + B_2, & \text{if } x_1 + x_3 \geq 0 \text{ and } -x_1 + x_3 \leq 0; \\ A_2\mathbf{x} + B_4, & \text{if } x_1 + x_3 > 0 \text{ and } -x_1 + x_3 > 0; \\ A_1\mathbf{x} + B_3, & \text{otherwise;} \end{cases} \qquad (2.76)$$

with A_1 and A_2 given by (2.45)

The modified system is as follows, consider the systems proposed in (2.76) with the vectors B_i, $i = 1, \ldots, 4$, given as follows:

$$B_1 = -B_4 = \begin{pmatrix} 0.1 \\ 2 \\ 0.3 \end{pmatrix} + k \begin{pmatrix} 0.2 \\ 4 \\ 0.2 \end{pmatrix} \text{ and } B_2 = -B_3 = \begin{pmatrix} 0.35 \\ -2 \\ 0.075 \end{pmatrix} + k \begin{pmatrix} 0.25 \\ -4 \\ 0.15 \end{pmatrix},$$

$$(2.77)$$

where $0 \leq k \in \mathbb{R}$. Figure 2.37(a) shows the bistable system given by $k = 0.5$, when $k \approx 0$ the system presents four scrolls in the attractor, i.e., a quadruple-scroll attractor with square shape which will be called *square chaotic attractor*.

When $k = 0$ the two heteroclinic orbits collide at $(0, 0, 0)^T$ and then the unstable manifolds of $\mathbf{x}_1^*$ and $\mathbf{x}_4^*$ are joined to the stable manifold of $\mathbf{x}_2^*$. In the Fig. 2.37(b) it is shown the projection of the square chaotic attractor for $k = 0$ and initial condition $\mathbf{x}(0) = (0.8, 0.2, -0.8)^T$ onto the plane (x_1, x_3).

For $k = 0$ the system is mono-stable, so no matter where an initial condition starts if it belongs to the basin of attraction, then the trajectory converges to the attractor around the intersection of the switching planes, otherwise the trajectory grows to infinite. This can be seen in Fig. 2.37(d) that shows the new basin of attraction corresponding to the attractor.

2.7 Hyperchaotic systems

An attractor with two positive Lyapunov exponents is kwnown as hyperchaotic. One of the first hyperchaotic attractors was reported in [Rossler (1979)], the attractor appears in a five-dimensional system described as follows:

$$\begin{aligned} \dot{x} &\ -y - z \\ \dot{y} &\ x + 0.25y + w \\ \dot{z} &\ 3 + xz \\ \dot{w} &\ -0.5z + 0.5w. \end{aligned} \qquad (2.78)$$

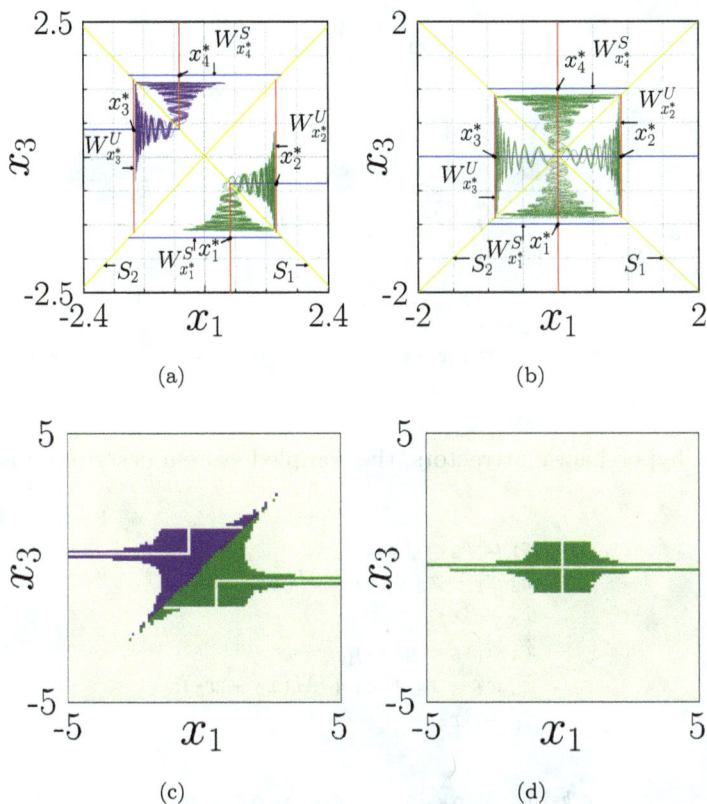

Fig. 2.37   In (a) the two attractors of the bistable system given by (2.45), (2.47), and (2.77) with $\gamma_1 = 0.2$, $\gamma_2 = 0.25$, $\sigma_1 = 0.2$, $\sigma_2 = 0.15$, and $\omega = 4$ for the initial conditions $\mathbf{x}(0) = (-0.8, -0.2, 0.8)^T$ (purple) and $\mathbf{x}(0) = (0.8, 0.2, -0.8)^T$. In (b) the chaotic four-scroll attractor of the system given by (2.45), (2.47), and (2.48) for the same parameters and $k = 0.0$ for the initial condition $\mathbf{x}(0) = (0.8, 0.2, -0.8)^T$. In (c) and (d) the estimated basins of attraction of the bistable and monostable systems, respectively. The dot colors are in concordance with the color of the attractor.

The computed Lyapunov exponents for this attractor are $\lambda_1 = 0.011$, $\lambda_2 = 0.02$, $\lambda_3 = -24.9$ and $\lambda_1 =$. Two projections of the hyperchaotic attractor are shown in Fig. 2.38.

A known approach to generate hyperchaotic attractors is to couple two three-dimensional systems with a chaotic attractor. As example, in [Cafagna and Grassi (2003)] two moddified Chua systems were coupled to

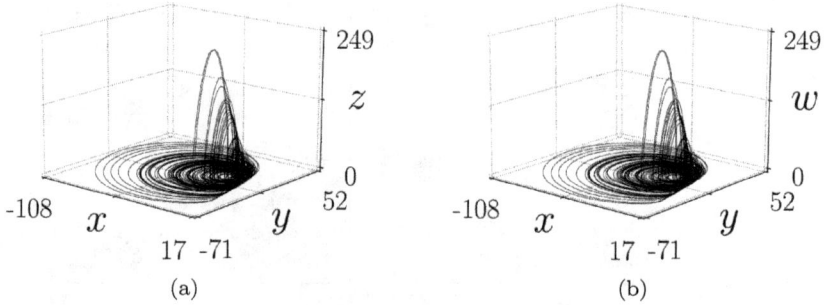

Fig. 2.38 Hyperchaotic attractor exhibited by the system (2.78) projected onto (a) (x, y, z) and (b) (x, y, w).

generate hyperchaotic attractors, the compled system description is as follows:

$$
\begin{aligned}
\dot{x}_1 &\; \alpha[x_2 - f(x_1)], \\
\dot{x}_2 &\; x_1 - x_2 + x_3 + H(x_5 - x_2), \\
\dot{x}_3 &\; -\beta x_2, \\
\dot{x}_4 &\; \alpha[x_5 - g(x_4)], \\
\dot{x}_5 &\; x_4 - x_5 + x_6 + M(x_2 - x_5), \\
\dot{x}_6 &\; -\beta x_5,
\end{aligned}
\tag{2.79}
$$

$$
f(x_1) = \begin{cases}
\frac{b*\pi}{2a}(x_1 - 2ac_1) & \text{if } x_1 \geq 2ac_1, \\
-b/sin(\frac{\pi x_1}{2a} + d_1) & \text{if } -2ac_1 < x_1 < 2ac_1, \\
\frac{b*\pi}{2a}(x_1 + 2ac_1) & \text{if } x_1 \leq -2ac_1,
\end{cases}
\tag{2.80}
$$

$$
g(x_4) = \begin{cases}
\frac{b*\pi}{2a}(x_4 - 2ac_2) & \text{if } x_4 \geq 2ac_2, \\
-b/sin(\frac{\pi x_4}{2a} + d_2) & \text{if } -2ac_2 < x_4 < 2ac_2, \\
\frac{b*\pi}{2a}(x_4 + 2ac_2) & \text{if } x_4 \leq -2ac_2.
\end{cases}
\tag{2.81}
$$

For $\alpha = 10.814$, $\beta = 14$, $a = 1.3$, $b = 0.11$, $c_1 = 7$, $c_2 = 3$, $d_1 = d_2 = 0$ the system exhibits a hyperchaotic attractor which is shown in the Fig. 2.39. The reported Lyapunov exponents for this attractor are $\lambda_1 = 0.340$, $\lambda_2 = 0.179$, $\lambda_3 = 0$, $\lambda_4 = -0.165$, $\lambda_5 = -1.374$ and $\lambda_1 = -1.833$.

In [Escalante-González and Campos (2020a)] a similar approach is used to generate a hidden hyperchaotic attractor. The base three-dimensional

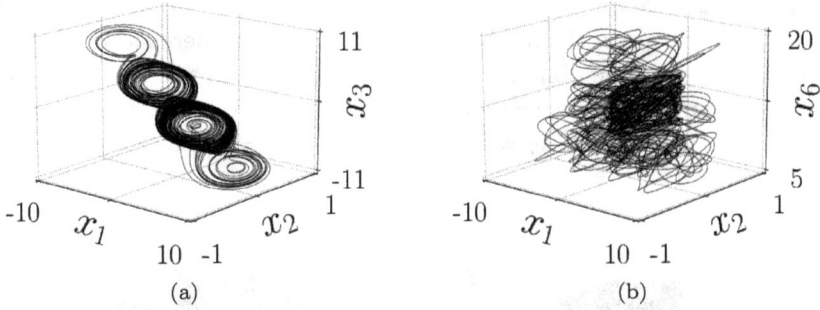

Fig. 2.39 Hyperchaotic attractor exhibited by (2.79), (2.80) and (2.81) projected onto (a) (x_1, x_2, x_3) and (b) (x_1, x_2, x_6).

system is as follows:

$$\begin{aligned}
\dot{x} &= by, \\
\dot{y} &= -b(x - f(X)) + ay, \\
\dot{z} &= -|y|v(z - f(X)) + d,
\end{aligned} \tag{2.82}$$

$$f(X) = \alpha Q((x + z + 2\gamma) + (\gamma - \alpha)Q((x + z)) + \alpha Q((x + z - 2\gamma)) \tag{2.83}$$

with $a, b, v > 0$ and $\alpha, \gamma \geq 0$ and Q one of the following three functions:

$$H_1(x) = \tanh(kx), \tag{2.84}$$

$$H_2(x) = \begin{cases} 1, & \text{if } x \geq 0, \\ -1, & \text{if } x < 0, \end{cases} \tag{2.85}$$

or

$$H_3(x) = \begin{cases} -1, & \text{if } \quad x < -\frac{1}{k}, \\ k(x), & \text{if } -\frac{2}{k} \leq x \leq \frac{1}{k}, \\ 1, & \text{if } \quad x < \frac{1}{k}, \end{cases} \tag{2.86}$$

where $k \in \mathbb{R}$.

Then, three identical subsystems $\dot{X}_1$, $\dot{X}_2$ and $\dot{X}_3$ were coupled by modifying the $\dot{z}$ equation as follows:

$$\begin{aligned}
\dot{z}_1 &= -x_1^2 v(z_1 - f(X_1) - k_1 f(X_3)) + d, \\
\dot{z}_2 &= -x_2^2 v(z_2 - f(X_2) - k_2 f(X_1)) + d, \\
\dot{z}_3 &= -x_3^2 v(z_3 - f(X_2) - k_3 f(X_2)) + d.
\end{aligned} \tag{2.87}$$

As an example, consider the parameters $a = 1$, $b = 10$, $v = 5$, $k = 10$, $m = 1$, $p = 2$, $d = 0.1$ and $k_i = 0.1$, the hidden hyperchaotic attractor exhibited by the nine-dimensional system is shown in the Fig. 2.40. The reported Lyapunov exponents for this attractor are $\lambda_1 = 1.229$, $\lambda_2 = 1.124$, $\lambda_3 = 0.045$, $\lambda_4 = 1.029$, $\lambda_5 = 0$, $\lambda_6 = -0.044$, $\lambda_7 = -5.323$, $\lambda_8 = -6.362$ and $\lambda_9 = -7517$.

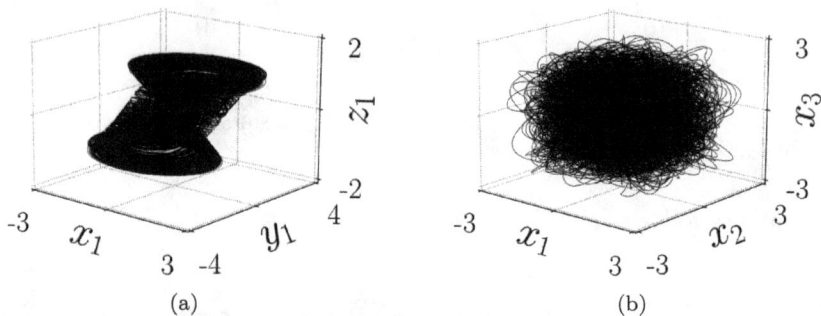

(a)   (b)

Fig. 2.40   Hyperchaotic attractor exhibited by (2.82), (2.83) and (2.84) projected onto (a) (x_1, x_2, x_3) and (b) (x_1, x_2, x_6).

Chapter 3

Systems with hidden attractors

3.1 Systems without equilibria

According to [Leonov *et al.* (2011b)], there exists two classes of attractors, the first one is the self-excited attractors which are said to be excited due to unstable equilibria. This means that the basin of attraction intersects with an arbitrarily small open neighborhood of the equilibria [Dudkowski *et al.* (2016)]. The majority of studied attractors belongs to this class, for instance, the Lorenz and Rössler attractors are self excited.

If a system has a self excited attractor, it can be found by numerical means selecting some initial conditions in the unstable manifold close to an unstable equilibrium point and computing the trajectories. As example, consider the system $\dot{X} = (\dot{x}, \dot{y}, \dot{z})^T = f(X)$ whose description is:

$$\begin{aligned}
\dot{x} &= by + a(x - \mathrm{sgn}(x - z)), \\
\dot{y} &= ay + b(\mathrm{sgn}(x - z) - x), \\
\dot{z} &= c(z + \mathrm{sgn}(x - z)),
\end{aligned} \qquad (3.1)$$

where $a = 0.2$, $b = 5$ and $c = -1$. The equilibria is located at $X_1^* = (-1, 0, 1)^T$, $X_2^* = (0, 0, 0)^T$ and $X_3^* = (1, 0, -1)^T$ and the local unstable manifolds are given by $W_i^U = X_i^* + k_1 e_1 + k_2 e_2$, where $k_1, k_2 \in \mathbb{R}$, $e_1 = (1, 0, 0)^T$ and $e_2 = (0, 1, 0)$.

Let us chose some initial conditions of the form $X_0 = (-1, \epsilon, 1)^T \in W_1^U$ with $\epsilon \leq 0.1$. The solution $X(t)$ for $t \in [5000, 5100]$ remains close to the equilibria for all the t, revealing the existence of an attractor. In Fig. 3.1, it is shown the solution for $X_0 = (-1, 0.1, 1)^T$.

The second class of attractors are called hidden attractors and are those not excited by unstable equilibria, i.e., we can find a neighborhoods around

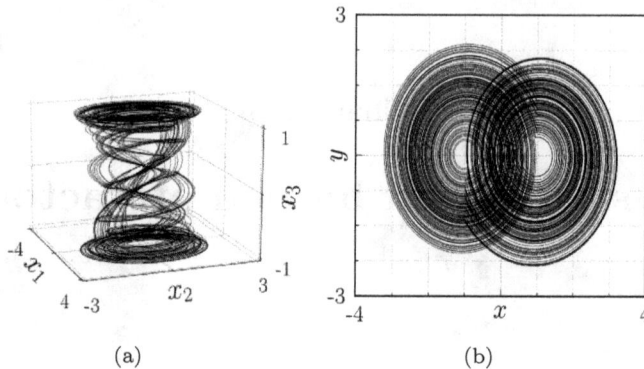

(a)   (b)

Fig. 3.1   Attractor of the system (3.1) for $X_0 = (-1, 0.1, 1)^T$ and $t \in [5000, 5300]$ in (a) $x - y - z$ and (b) its projection onto the plane (x, y).

each equilibria such that there is no intersection of these neighborhoods with the basin of attraction.

An attractor in a vector without equilibria is also called a non-self-excited attractors.

The location of hidden attractors in a systems with equilibria as well as non-self-excited attractors in a system without equilibria is not as simple as in the case of self-excited attractors and thus it is interesting to understand the dynamics behind its existence. However, it is also relevant to understand them from a practical point of view. For example, in a control systems it is important to be aware of their existence since they could lead us to undesired behavior. An example of this unexpected behavior is a well known incident with the the aircraft YF-22 Boeing in 1992 [Andrievsky *et al.* (2013)].

In [Leonov *et al.* (2011a)] it is discussed a special analytical-numerical algorithm for the localization of hidden attractors in a system.

One approach that can be used to find the description of a system with bounded flows is the use of numerical simulations of candidate equations. In [Sprott (1994)] a numerical search of three-dimensional chaotic flows whose descriptions had one or two nonlinearities was performed. From the 19 founded flows called Sprott cases, one is specially interesting, the Sprott case A, which is a system without equilibria with a conservative flow with chaotic behavior. The system Sprott case A is a special case of the Nose-Hoover oscillator [Hoover (1995)] and it exhibits a chaotic solution for the initial condition $X = (0, 5, 0)^T$. After this system was reported, there were an increment in the search of new systems without equilibria with chaotic flows and hidden attractors. The description of system Sprott case A is the following:

$$\dot{x} = y,$$
$$\dot{y} = -x + yz, \tag{3.2}$$
$$\dot{z} = 1 - y^2.$$

The reported Lyapunov exponents are $\{0.014, 0, -0.014\}$. The plot of the solution projected onto the plane $x - y$ for $X = (0, 5, 0)^T$ and $t \in [0, 2000]$ is shown in the Fig. 3.2.

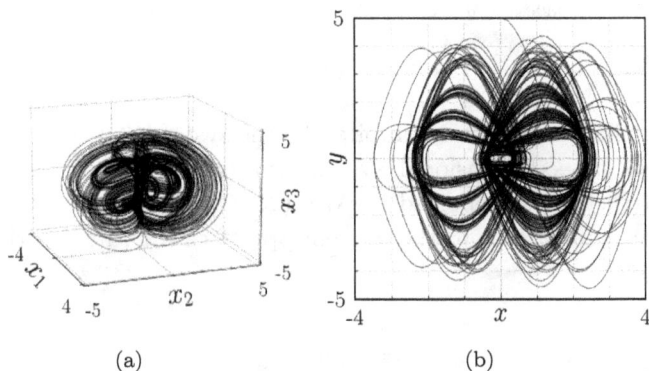

(a)   (b)

Fig. 3.2 Solution of the system given by (3.2) for $X = (0, 5, 0)^T$ and $t \in [0, 2000]$ in (a) the space $x - y - z$ and (b) its projection onto the plane $x - y$.

In [Wang and Chen (2013)] a methodology to construct a system with any number of equilibria and a chaotic attractor was introduced. This work also introduced a system without equilibria with a chaotic attractor. Its description is the following:

$$\dot{x} = y,$$
$$\dot{y} = z, \tag{3.3}$$
$$\dot{z} = -y + 3y^2 - x^2 - xz + a.$$

The system can present two symmetrical equilibria for $a > 0$, one equilibrium point at the origin for $a = 0$ or not equilibria for $a < 0$. In Fig. 3.3 it is

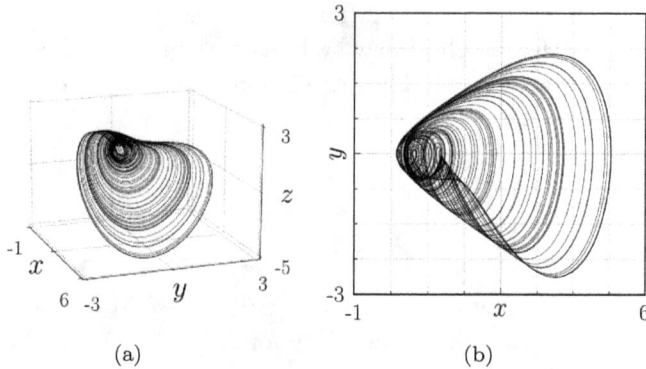

Fig. 3.3   Hidden attractor of the system given by (3.3) for $X = (5, 0.1, 0)^T$ and $t \in [0, 500]$ in (a) the space $x - y - z$ and (b) its projection onto the plane $x - y$.

shown the chaotic attractor exhibited by the system (3.3) when $a = -0.05$.

Another three-dimensional system without equilibria with a hidden chaotic attractor was reported in [Wei (2011)]. The system resulted from adding a constant term to the System Sprott case D, its description is as follows:

$$\dot{x} = -y,$$
$$\dot{y} = cx + z, \qquad\qquad (3.4)$$
$$\dot{z} = ay^2 + xz - d,$$

where $a, c, d \in \mathbb{R}$. The system is identical to the Sprott case D when $a = 3$, $c = 1$ and $d = 0$. The particular case reported in [Wei (2011)] considers the parameters $a = 2$, $c = 1$ and $d = 035$ along with the initial condition $X = (-1.6, 0.82, 1.9)^T$. The Lyapunov exponents for this parameters are reported as $\{0.0793, 0, -1.5034\}$. The hidden attractor is shown in the Fig. 3.4.

In [Jafari *et al.* (2013)] a systematic search was performed based on three methods to produce a catalog of seventeen three-dimensional flows with quadratic nonlinearities and no equilibria at all. The found systems were called NE systems. However, three systems in the catalog were already known from [Sprott (1994); Wei (2011); Wang and Chen (2013)].

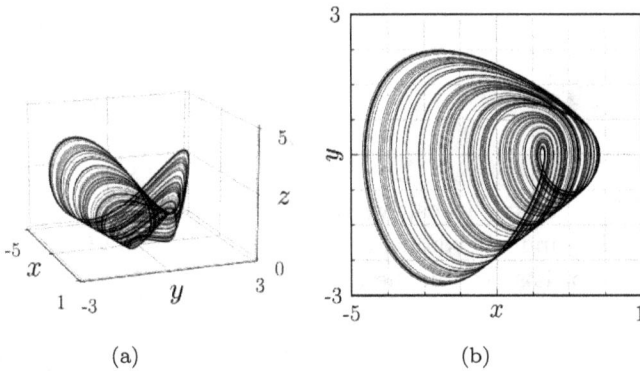

Fig. 3.4 Hidden attractor of the system given by (3.4) for $a = 2$, $c = 1$, $d = 035$, $X = (-1.6, 0.82, 1.9)^T$ and $t \in [0, 500]$ in (a) the space $x - y - z$ and (b) its projection onto the plane $x - y$.

To illustrate these systems consider the NE6 which is described as follows:

$$\begin{aligned}
\dot{x} &= y, \\
\dot{y} &= z, \\
\dot{z} &= -y - xz - yz - a.
\end{aligned} \qquad (3.5)$$

In order to find the equilibria let us equate to zero the vector field:

$$\begin{aligned}
0 &= y, \\
0 &= z, \\
0 &= -y - xz - yz - a,
\end{aligned} \qquad (3.6)$$

thus,

$$0 = -a, \qquad (3.7)$$

which implies there is no equilibria. The reported Lyapunov exponents for $a = 0.75$ are $\{0.0280, 0, -3.434\}$. The hidden attractor is shown in the Fig. 3.5. As a second example from this catalog, consider the system NE14:

$$\begin{aligned}
\dot{x} &= y, \\
\dot{y} &= z, \\
\dot{z} &= x^2 - y^2 + 2xz + yz + a.
\end{aligned} \qquad (3.8)$$

In order to find the equilibria let us equate to zero the vector field:

$$\begin{aligned}
0 &= y, \\
0 &= z, \\
0 &= x^2 - y^2 + 2xz + yz + a,
\end{aligned} \qquad (3.9)$$

thus,

$$0 = x^2 + a, \qquad (3.10)$$

$$x = \pm\sqrt{-a}, \qquad (3.11)$$

which implies that for a positive value of a the system has no equilibria. The reported Lyapunov exponents for $a = 1$ are $\{0.0532, 0, -11.8580\}$. The hidden attractor is shown in the Fig. 3.6.

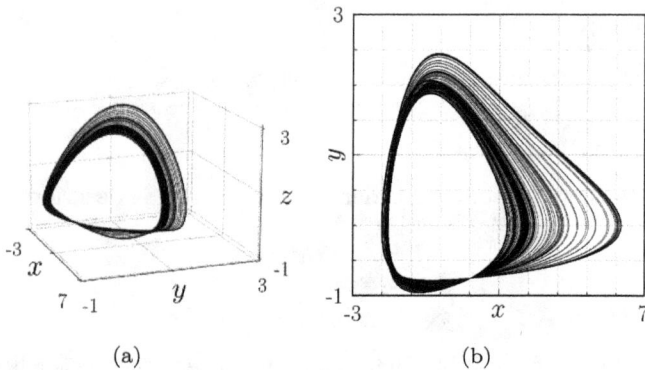

(a)   (b)

Fig. 3.5   Hidden attractor of the system given by (3.5) for $a = 0.75$ and $X = (0, 3, -0.1)^T$ and $t \in [2000, 4000]$ in (a) the space $x - y - z$ and (b) its projection onto the plane $x - y$.

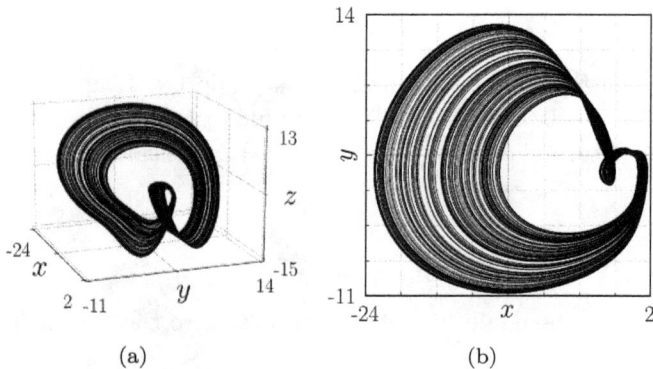

(a)   (b)

Fig. 3.6   Hidden attractor of the system given by (3.8) for $a = 1$ and $X = (1, 0, -4)^T$ and $t \in [2000, 4000]$ in (a) the space $x - y - z$ and (b) its projection onto the plane $x - y$.

Systems without equilibria with higher dimension and hidden attractors have also been reported. One widely known example is the system introduced in [Li *et al.* (2014)], which is also the first reported system with a globally attracting hyperchaotic hidden attractor in a system without equilibria. The reported system was the result of adding a linear feedback from an additional variable to the diffusionless Lorenz systems and the approximation of the quadratic terms by means of sign and absolute value functions. The system description is as follows:

$$\begin{aligned}
\dot{x} &= y - x, \\
\dot{y} &= -z\text{sgn}(x) + u, \\
\dot{z} &= |x| - a, \\
\dot{u} &= -by,
\end{aligned}$$

(3.12)

where a is the amplitude parameter that determines the size of the attractor. The system is hyperchaotic for $b \in [0.1, 0.3]$. When $a = 1$ and $b = 0.25$ the hidden attractor is hyperchaotic with Lyapunov exponents $\{0.064, 0.033, 0, -1.098\}$. The attractor is shown in the Fig. 3.7.

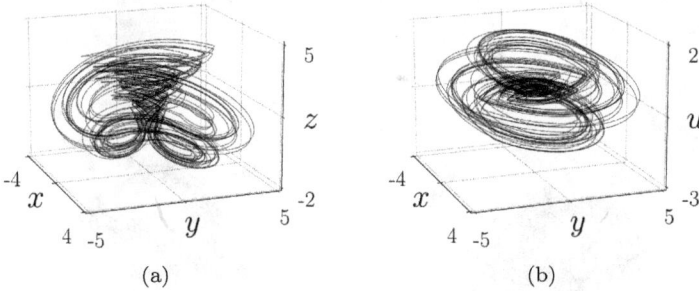

Fig. 3.7 Projection of the hidden hyperchaotic attractor of the system given by (3.12) for $a = 1$, $b = 0.25$, $X = (0.1, 0.1, 0.1)^T$ and $t \in [500, 1000]$ in (a) $x - y - z$ and (b) $x - y - u$.

As seen from previous examples, to find systems with hidden attractors numerical search can be performed on certain type of vector fields that are expected to present bounded flows. Another consist in the construction of of the vector field by a PWL description. Working with PWL systems is usually easier than working with system descriptions that contains quadratic or cubic nonlinearities.

A PWL three-dimensional system could be defined as follows:

$$\dot{X} = A_i X + B_i, \quad X \in P_i,$$

(3.13)

where $X = (x_1, x_2, x_3)^T \in \mathbb{R}^3$ is the state vector, $A_i = \{\alpha_{lm}\} \in \mathbb{R}^{3 \times 3}$ is a linear operator and $B_i \in \mathbb{R}$ is a constant vector.

A PWL system has no equilibria if for each atom P_i, one of the following cases is true:

(1) $A_i + B_i \neq 0$
(2) $-A_i^{-1} B_i \notin P_i$

For instance, consider the following system which exhibits a hidden attractor [Escalante-González and Campos-Cantón (2017)]:

$$\dot{X} = \begin{cases} AX + B_1, & \text{if } x_1 < 0, \\ AX + B_2, & \text{if } x_1 \geq 0, \end{cases} \tag{3.14}$$

where

$$A = \begin{bmatrix} 0 & 1 & 1 \\ 0 & -0.5 & 3 \\ 0 & -3 & -0.5 \end{bmatrix}, \quad B_1 = \begin{bmatrix} 0.5 \\ 0 \\ 0 \end{bmatrix}, \quad B_2 = \begin{bmatrix} -1 \\ 10 \\ 0 \end{bmatrix}. \tag{3.15}$$

The partition has two atoms, P_1 and P_2. For each atom $A + B_i \neq 0$ and thus the system has no equilibria. In the Figure 3.8 is shown the hidden attractor in $x_1 - x_2 - x_3$ and its projection onto $x_1 - x_2$ for $X_0 = (0, 0, 0)^T$ and $t \in [100, 1100]$.

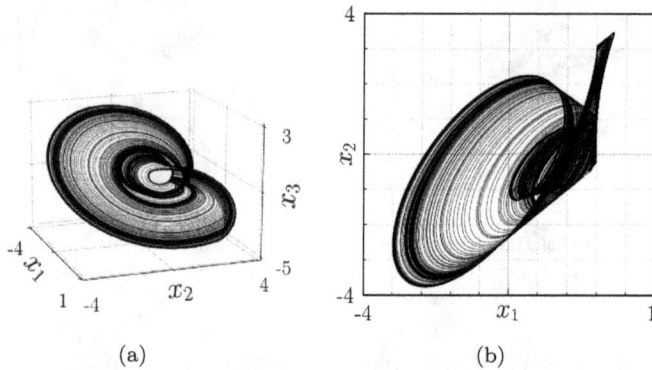

(a)   (b)

Fig. 3.8   Hidden attractor of the system given by (3.14) and (3.15) for $X = (0, 0, 0)^T$ and $t \in [100, 1100]$ in (a) $x_1 - x_2 - x_3$ and (b) its projection onto the plane $x_1 - x_2$.

Consider now the following systems which also exhibits a hidden attractor [Wei *et al.* (2021)]:

$$\dot{x} = \begin{cases} A_1 X + B_1, & C^T x \leq 1, \\ A_2 x + B_2, & C^T x > 1, \end{cases} \tag{3.16}$$

where $C = (-1/3, -1/3, 1)^T$ and

$$A_1 = \begin{bmatrix} -0.5 & -3 & 0 \\ 3 & -0.5 & 0 \\ 0 & 0 & -0.1 \end{bmatrix}, \quad A_2 = \begin{bmatrix} -0.5 & -3 & 0 \\ 3 & -0.5 & 0 \\ 0 & -0.2 & -0.1 \end{bmatrix}, \quad (3.17)$$

$$B_1 = \begin{bmatrix} 0 \\ 0 \\ 0.5 \end{bmatrix}, \quad B_2 = \begin{bmatrix} -10 \\ 0 \\ -4 \end{bmatrix}. \quad (3.18)$$

The partition has also two atoms, P_1 and P_2. For each atom $-A_i^{-1}B_i \notin P_i$ and thus the system has no equilibria. Its hidden attractor for $X_0 = (-1, 0.1, 1)^T$ and $t \in [100, 600]$ is shown in the Fig. 3.9.

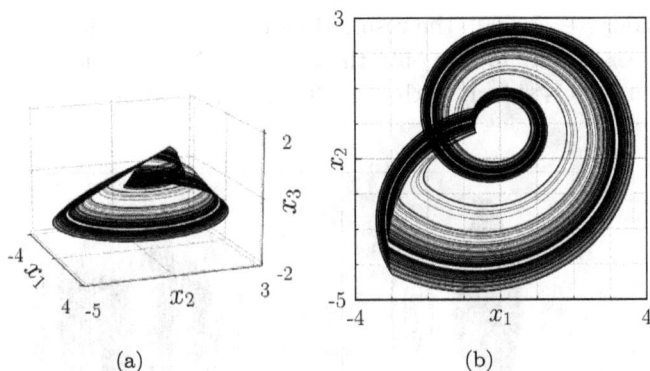

Fig. 3.9 (a) Hidden attractor of the system given by (3.16), (3.17) and (3.18) in $x_1 - x_2 - x_3$ and (b) its projection onto the plane $x_1 - x_2$.

It is important to note that the absence of equilibria does not imply that there exist an attractor and thus further considerations are needed which has lead to the publication of different methodologies which usually depends on numerical test.

3.2 Hidden scroll attractors

As seen in the previous chapter, there exist several works that have addressed the topic of scroll attractors, its generation, design and applications. The majority of design approaches are based on the fact that a scroll can be formed around one or more equilibrium points. In the case of hidden scroll attractors, those techniques can be applied directly.

In [Hu *et al.* (2016)] two three-dimensional systems which exhibits multi-scroll hidden attractors are constructed by adding nonlinear functions into the system known as Sprott case A reported in [Sprott (1994)]. The first system proposed in [Hu *et al.* (2016)] is the following:

$$\begin{aligned}
\dot{x} &= y, \\
\dot{y} &= -x + yz - a\sin(2\pi bx), \\
\dot{z} &= 1 - y^2,
\end{aligned} \tag{3.19}$$

where $a = 25$ and $b = 1$. The system has no equilibria and for the initial condition $X_0 = (0, 0.1, 0)^T$ it exhibits a multi-scroll hidden attractor whose scroll number depends on the transient time, i.e. for a longer simulation time the size the number of scrolls plotted is increased. To see the effect of time, consider $t \in [0, 500]$, the resulted plot is shown in the Fig. 3.10(a), as it can be seen, there appear only three scrolls. Now consider $t \in [0, 1000]$, the resulted plot is shown in the Fig. 3.10(b), now appear six scrolls.

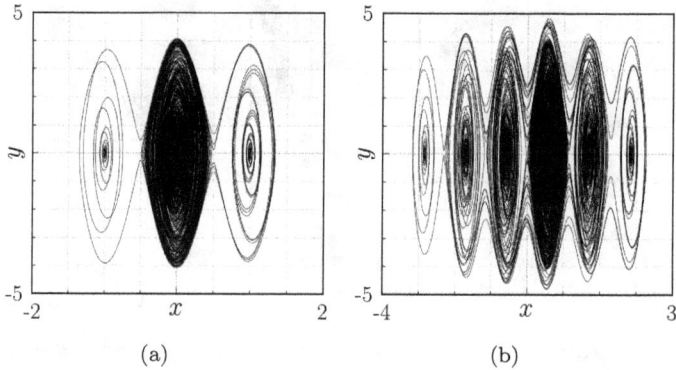

(a)   (b)

Fig. 3.10   (a) Hidden attractor of the system given by (3.19) with $a = 25$, $b = 1$, $X_0 = (0, 0.1, 0)^T$ and $t \in [0, 500]$ projected onto the plane $x - y$ and (b) for $t \in [0, 1000]$.

The second proposed class of systems modifies the nonlinearity in order to limit the scrolls along the x axis to the subset $x \in [c, d]$. Its description is as follows:

$$\begin{aligned}
\dot{x} &= y, \\
\dot{y} &= -x + yz - af(x), \\
\dot{z} &= 1 - y^2,
\end{aligned} \tag{3.20}$$

where

$$f(x) = \sin(2\pi bx)(sgn(x-c) - sgn(x-d)) + x(2 - sgn(x-c) + sgn(x-d)). \quad (3.21)$$

To see the difference with the previous case, consider $a = 25$, $b = 1$, $c = -1.5$, $d = 1.5$ and $t \in [0, 1000]$, the resulted plot is shown in the Fig. 3.11(a). Now consider the same parameters but with $t \in [0, 3000]$, as it can be seen, the number of scrolls remains unchanged, the resulted plot is shown in the Fig. 3.11(b). Now $a = 25$, $b = 1$, $c = -3.5$, $d = 3.5$ with $t \in [0, 1000]$ and $t \in [0, 3000]$, the number of scrolls also remains unchanged and the resulted plots are shown in the Figs. 3.11(c) and 3.11(d), respectively.

In [Escalante-González *et al.* (2017)] a different class of systems with hidden multi-scroll attractors was reported. The idea behind this class of systems is the "saddle-focus like" equilibrium point.

To introduce this idea of the "saddle-focus like" let consider the following system

$$\dot{X} = AX \text{ with } X = (x_1, x_2, x_3)^T, \quad (3.22)$$

where

$$A = \begin{pmatrix} m & -n & 0 \\ n & m & 0 \\ 0 & 0 & 0 \end{pmatrix}, \quad A = (a_1, a_2, a_3), \quad (3.23)$$

such that $m, n > 0$. The linear operator A has no inverse and the system has an infinite number of equilibria. However if the system is modified as follows

$$\dot{X} = AX + V, \quad (3.24)$$

where

$$V = \begin{pmatrix} 0 \\ 0 \\ v \end{pmatrix} \quad (3.25)$$

with $v \neq 0$ and $v \in \mathbb{R}$ then the system has no equilibria at all. Now consider the following system

$$\dot{x} = \begin{cases} AX + V + W, & \text{if } X < C, x_1 < 0; \\ AX + V - W, & \text{if } X < C, x_1 \geq 0; \\ AX + W, & \text{if } X = C, x_1 \leq 0; \\ AX - W, & \text{if } X = C, x_1 \geq 0; \\ AX - V + W, & \text{if } X > C, x_1 < 0; \\ AX - V - W, & \text{if } X > C; x_1 \geq 0; \end{cases} \quad (3.26)$$

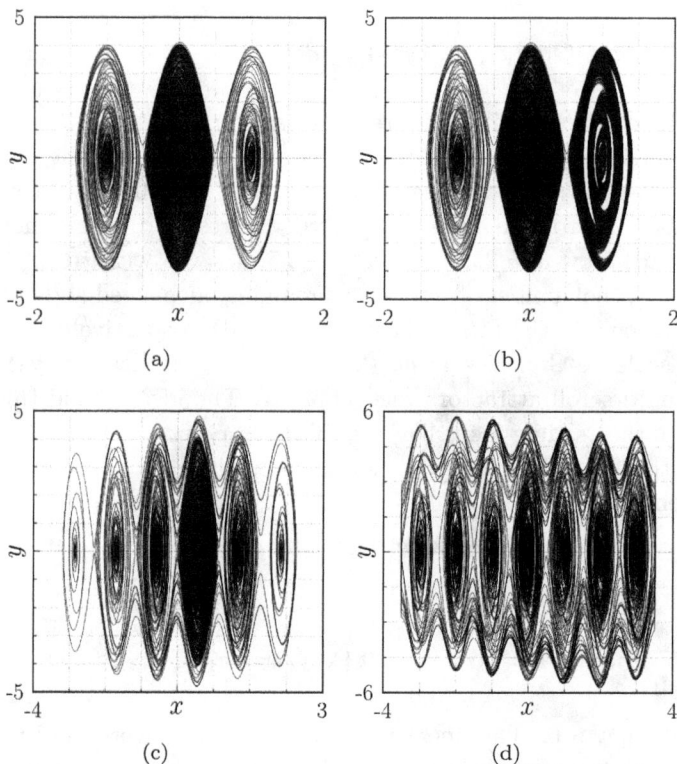

Fig. 3.11　Hidden attractor of the system given by (3.20) with $a = 25$, $b = 1$, $c = -1.5$, $d = 1.5$ and $X_0 = (0, 0.1, 0)^T$ projected onto the plane $x - y$ for (a) $t \in [0, 1000]$ and (b) and $t \in [0, 1000]$. Hidden attractor of the system given by (3.20) with $a = 25$, $b = 1$, $c = -3.5$, $d = 3.5$ and $X_0 = (0, 0.1, 0)^T$ projected onto the plane $x - y$ for (c) $t \in [0, 1000]$ and (d) $t \in [0, 3000]$.

where $C = [0, 0, \tau]X$ with $\tau \in \mathbb{R}$ and

$$W = k_1 a_1 + k_2 a_2, \tag{3.27}$$

$$W = k_1 a_1 + k_2 a_2, \tag{3.28}$$

with k_1, $k_2 \in \mathbb{R}$ such that the equilibria of $\dot{X} = AX + W$ has its first element $x_1 > 0$. Then, the system given by (3.26) resembles a saddle focus equilibrium point at $(0, 0, \tau)^T$ with an "unstable manifold like" at the switching plane $S = \{X \in \mathbb{R} : x_3 = \tau\}$ whose normal vector is given

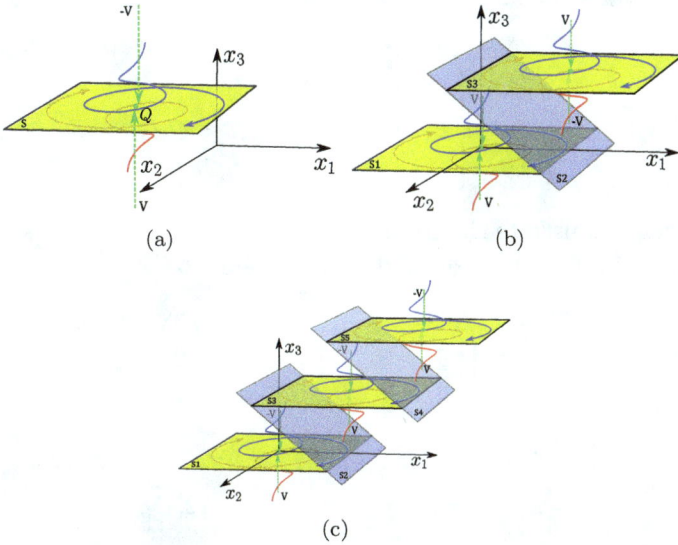

Fig. 3.12   (a) Illustration of the idea of "saddle-focus like". (b) Construction idea for a double-scroll attractor. (c) Construction idea for a triple-scroll attractor.

by $n = (0, 0, 1)^T$. The idea is shown in the Fig. 3.12. It is well known that systems with multi-scroll attractors can be designed by using saddle-foci and thus, the correct location of this subsystems with the "saddle-focus like" behaviors can lead to a system without equilibria with a hidden multi-scroll attractor. For instance, a double scroll can be generated by using two subsystems of the form given by (3.26) $F_1(x)$ and $F_3(x)$ and defining an additional switching surface S_2 with normal vector n_2. The idea is shown in the Fig. 3.12(b).

By defining $C_2 = n_2^T X$ the system with a hidden double-scroll attractor could be described as

$$\dot{x} = \begin{cases} F_1(X), & \text{if } X < C_2; \\ F_3(X), & \text{if } X \geq C_2. \end{cases} \tag{3.29}$$

It is also possible to extend the idea to more subsystems in order to generate more scrolls just by adding switching surfaces S_i with their respective C_i and n_i. For example if a system with three scrolls is desired the idea for the construction would be similar to the shown in the Fig. 3.12(b) where two additional switching surfaces S_2 and S_4 are needed. Its description would be:

$$\dot{x} = \begin{cases} F_1(X), & \text{if } X < C_2; \\ F_2(X), & \text{if } C_2 \geq X < C_4; \\ F_3(X), & \text{if } X \geq C_4. \end{cases} \tag{3.30}$$

In order to illustrate the construction idea let us consider first a system given by (3.22), (3.23), (3.26) and (3.29) where

$$A = \begin{pmatrix} 0.5 & -9 & 0 \\ 9 & 0.5 & 0 \\ 0 & 0 & 0 \end{pmatrix}, \quad V = \begin{pmatrix} 0 \\ 0 \\ 3 \end{pmatrix}, \tag{3.31}$$

with the PWL system:

$$\dot{X} = F_i(X), \quad i = 1, \ldots, 12, \tag{3.32}$$

The switching surfaces between subsystems and its normal vectors are:

$$\begin{aligned} S_1 &= \{x \in \mathbb{R}^3 : x_3 = 0\}, \ n_1 = x_3, \\ S_2 &= \{x \in \mathbb{R}^3 : x_1 + x_3/2 = 1\}, \ n_2 = x_1 + 0.5x_3, \\ S_3 &= \{x \in \mathbb{R}^3 : x_3 = 2, \ n_3 = x_3. \end{aligned} \tag{3.33}$$

Then the description is as follows:

$$\dot{X} = \begin{cases} AX + V + W_1, & \text{if } C_1 < 0, x_1 < 0; \\ AX + V + W_2, & \text{if } C_1 < 0, C_2 < 1, x_1 \geq 0; \\ AX + W_1, & \text{if } C_1 = 0, x_1 < 0; \\ AX + W_2, & \text{if } C_1 = 0, C_2 < 1, x_1 \geq 0; \\ AX - V + W_1, & \text{if } C_1 > 0, C_2 < 1, x_1 < 0; \\ AX - V + W_2, & \text{if } C_1 > 0, C_2 < 1, x_1 \geq 0; \\ AX + V + W_3, & \text{if } C_3 < 2, C_2 \geq 1, x_1 < 1; \\ AX + V + W_4, & \text{if } C_3 < 2, C_2 \geq 1, x_1 \geq 1; \\ AX + W_3, & \text{if } C_3 = 2, C_2 \geq 1, x_1 < 1; \\ AX + W_4, & \text{if } C_3 = 2, x_1 \geq 1; \\ AX - V + W_3, & \text{if } C_3 > 2, C_2 \geq 1, x_1 < 1; \\ AX - V + W_4, & \text{if } C_3 > 2, x_1 \geq 1. \end{cases} \tag{3.34}$$

Vectors W_i are as follow:

$$W_1 = -0.1a_1$$
$$W_2 = 0.1a_1$$
$$W_3 = -1.1a_1 \qquad\qquad (3.35)$$
$$W_4 = -0.9a_1.$$

The hidden double-scroll attractor with is shown in the Fig. 3.13(a) and its projection onto the plane $x_1 - x_2$ in the Fig. 3.13(b).

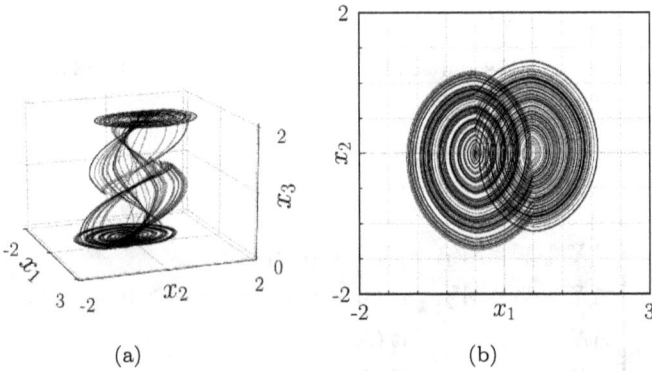

(a)　　　　　　　　(b)

Fig. 3.13　Hidden double-scroll attractor of the system given by (3.31), (3.33), (3.34) and (3.35) for the initial condition $(0,0,0)^T$ in (a) $x_1 - x_2 - x_3$ and its projection onto the planes (b) $x_1 - x_2$.

This system can be extended to a triple-scroll attractor by adding a subsystem of the form (3.26) and another switching surface S_4. Thus, all the switching surfaces and normal vectors are:

$$S_1 = \{x \in \mathbb{R}^3 : x_3 = 0\}, \; n_1 = x_3,$$
$$S_2 = \{x \in \mathbb{R}^3 : x_1 + x_3/2 = 1\}, \; n_2 = x_1 + 0.5x_3,$$
$$S_3 = \{x \in \mathbb{R}^3 : x_3 = 2, \; n_3 = x_3, \qquad\qquad (3.36)$$
$$S_4 = \{x \in \mathbb{R}^3 : x_1 + x_3/2 = 3\} \; n_4 = x_1 + 0.5x_3,$$
$$S_5 = \{x \in \mathbb{R}^3 : x_3 = 4, \; n_5 = x_3.$$

$$\dot{X} = F_i(X), i = 1, \ldots, 18. \qquad\qquad (3.37)$$

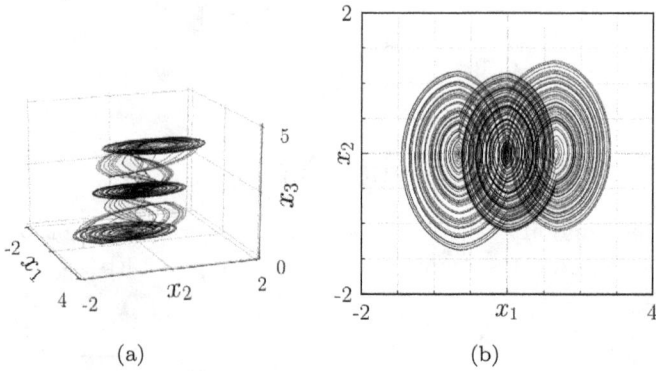

(a)   (b)

Fig. 3.14   Hidden double-scroll attractor of the system given by (3.31), (3.36), (3.37) and (3.39) for the initial condition $(0,0,0)^T$ in (a) $x_1 - x_2 - x_3$ and its projection onto the planes (b) $x_1 - x_2$.

$$\dot{X} = \begin{cases} AX + V + W_1, & \text{if } C_1 < 0, x_1 < 0; \\ AX + V + W_2, & \text{if } C_1 < 0, C_2 < 1, x_1 \geq 0; \\ AX + W_1, & \text{if } C_1 = 0, x_1 < 0; \\ AX + W_2, & \text{if } C_1 = 0, C_2 < 1, x_1 \geq 0; \\ AX - V + W_1, & \text{if } C_1 > 0, C_2 < 1, x_1 < 0; \\ AX - V + W_2, & \text{if } C_1 > 0, C_2 < 1, x_1 \geq 0; \\ AX + V + W_3, & \text{if } C_3 < 2, C_2 \geq 1, x_1 < 1; \\ AX + V + W_4, & \text{if } C_3 < 2, C_2 \geq 1, C_4 < 3, x_1 \geq 1; \\ AX + W_3, & \text{if } C_3 = 2, C_2 \geq 1, x_1 < 1; \\ AX + W_4, & \text{if } C_3 = 2, C_4 < 3, x_1 \geq 1; \\ AX - V + W_3, & \text{if } C_3 > 2, C_2 \geq 1, C_4 < 3, x_1 < 1; \\ AX - V + W_4, & \text{if } C_3 > 2, C_4 < 3, x_1 \geq 1 \\ AX + V + W_5, & \text{if } C_5 < 4, C_4 \geq 3, x_1 < 2; \\ AX + V + W_6, & \text{if } C_5 < 4, C_4 \geq 3, x_1 \geq 2; \\ AX + W_5, & \text{if } C_5 = 4, C_4 \geq 3, x_1 < 2; \\ AX + W_6, & \text{if } C_5 = 4, x_1 \geq 2; \\ AX - V + W_5, & \text{if } C_5 > 4, C_4 \geq 3, x_1 < 2; \\ AX - V + W_6, & \text{if } C_5 > 4, x_1 \geq 2. \end{cases} \qquad (3.38)$$

And now the vectors are:

$$
\begin{aligned}
W_1 &= -0.1a_1 \\
W_2 &= 0.1a_1 \\
W_3 &= -1.1a_1 \\
W_4 &= -0.9a_1 \\
W_5 &= -2.1a_1 \\
W_6 &= -1.9a_1.
\end{aligned}
\tag{3.39}
$$

The hidden triple-scroll attractor with is shown in the Fig. 3.14(a) and its projection onto the plane $x_1 - x_2$ in the Fig. 3.14(b).

Another class of hidden multi-scroll attractors was reported in [Escalante-González and Campos (2020b)], the class of systems is described as follows:

$$
\begin{aligned}
\dot{x} &= ax + b(y - f(X)), \\
\dot{y} &= -bx, \\
\dot{z} &= -x^2 v(z - f(X)) + d,
\end{aligned}
\tag{3.40}
$$

where $a, b, v > 0$, $b > a$ and $X = (x,\ y,\ z)^T$. The system has no equilibria when $d \neq 0$. $f(X)$ is a functional which defines the number of scrolls. Three types of functionals are proposed, the first one is:

$$
f(X) = \sum_{i=1}^{m} \frac{p}{2} \tanh \left(k(y + z + 2p(i-1) - p(m-1)) \right),
\tag{3.41}
$$

where the parameter $m \in \mathbb{Z}/\{0\}$ and the parameters $k, p \in \{X \in \mathbb{R} : X > 0\}$.

The second one is

$$
f(X) = \sum_{i=1}^{m} \frac{p}{2} H \left(y + z + 2p(i-1) - p(m-1) \right),
\tag{3.42}
$$

where

$$
H(x) = \begin{cases} 1, & \text{if } x \geq 0, \\ -1, & \text{if } x < 0, \end{cases}
\tag{3.43}
$$

and the same parameters m, k, p.

The third one is:

$$
f(X) = \sum_{i=1}^{m} \mathrm{sat}(y + z, 2p(i-1) - p(m-1)),
\tag{3.44}
$$

where

$$\text{sat}(x,c) = \begin{cases} -\frac{p}{2}, & \text{if} \quad x < c - \frac{p}{2k}, \\ k(x-c), & \text{if } c - \frac{p}{2k} \le x \le c + \frac{p}{2k}, \\ \frac{p}{2}, & \text{if} \quad x < c + \frac{p}{2k}, \end{cases} \tag{3.45}$$

and also the same parameters m, k, p.

All three functionals have in common a stair-like shape when plotted against $y + z$. In these plots, k and p determine the slope and the height between consecutive plateaus while m determines the number of plateaus. The number of scrolls in the attractor is given by $m + 1$. In the Fig. 3.15 are shown the plots of the functionals (3.41), (3.42) and (3.44) for $p = 2$, $k = 2$ and $m = 5$.

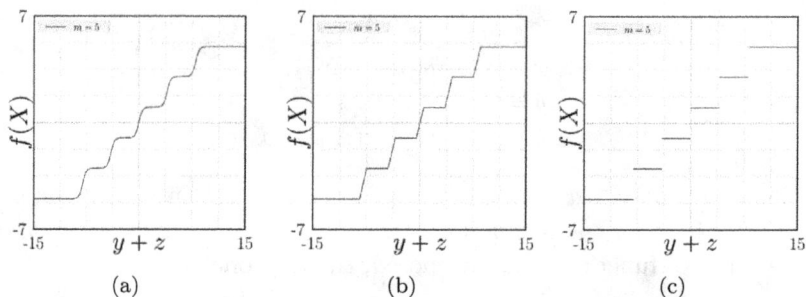

(a)   (b)   (c)

Fig. 3.15   Plot of the functionals (a) (3.41), (b) (3.42) and (c) (3.44) for $p = 2$, $k = 2$ and $m = 5$.

In order to illustrate the class of systems let us consider the parameters $a = 1, b = 10, v = 5, k = 10, m = 1, p = 2, d = 0.1$ with which the functionals are as follows:

$$\begin{aligned} f(X) &= \tanh(10(y+z)), \\ f(X) &= H(y+z), \\ f(X) &= \text{sat}(y+z,0). \end{aligned} \tag{3.46}$$

In the Fig. 3.16 is shown a comparison of the resulted attractor with the use of each functional. The attractor for the functional (3.41) is shown in the Fig. 3.16(b) and Fig. 3.16(a) in $x - y - z$ and $x - y$, respectively. The attractor for the functional (3.42) is shown in the Fig. 3.16(d) and Fig. 3.16(c) in $x - y - z$ and $x - y$, respectively. The attractor for the functional (3.44) is shown in the Fig. 3.16(f) and Fig. 3.16(e) in $x - y - z$ and $x - y$, respectively. As it can be seen from the Fig. 3.16, the resulting attractor is pretty similar, and thus, the selection would depend on which kind of vector field is desired.

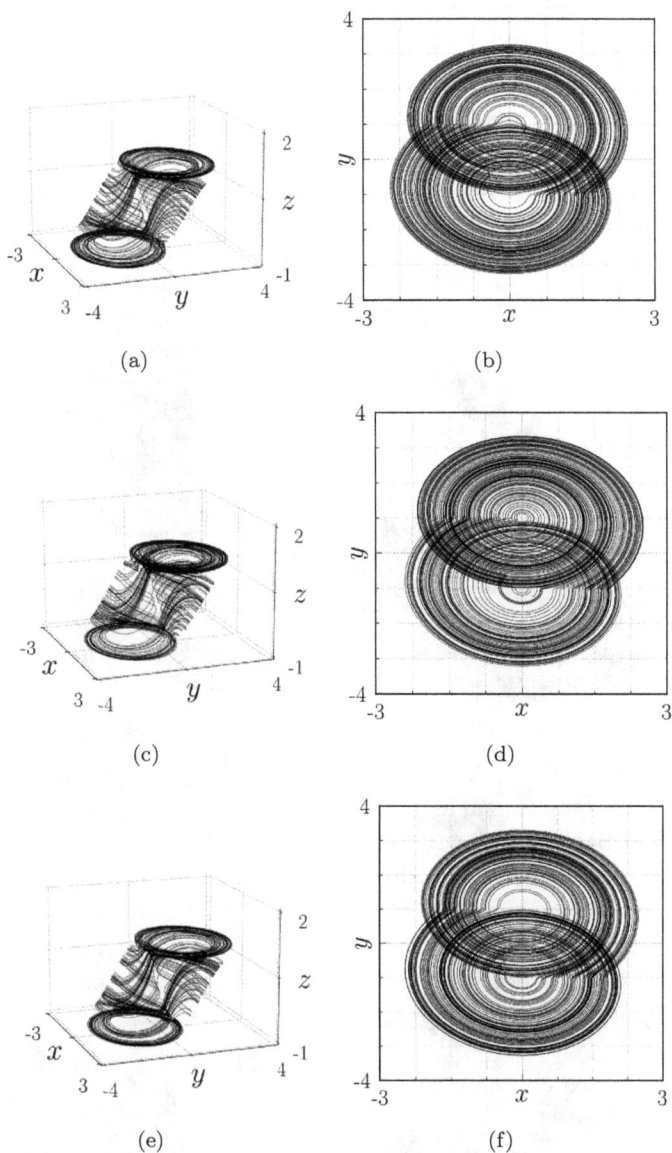

Fig. 3.16   Attractor for the functional (3.41) in (a) $x - y - z$ and its projection onto the plane (b) $x - y$. Attractor for the functional (3.42) in (b) $x - y - z$ and its projection onto the plane (c) $x - y$. Attractor for the functional (3.44) in (c) $x - y - z$ and its projection onto the plane (d) $x - y$.

In order to illustrate the generation of an attractor with a larger number of scrolls let us consider the functional given by (3.41) with the same parameters a, b, v, k, p, d and $m = 4$, thus the functional is:

$$f(X) = \tanh(10(y + z + 6)) + \tanh(10(y + z + 2))$$
$$+ \tanh(10(y + z - 2)) + \tanh(10(y + z - 6)). \qquad (3.47)$$

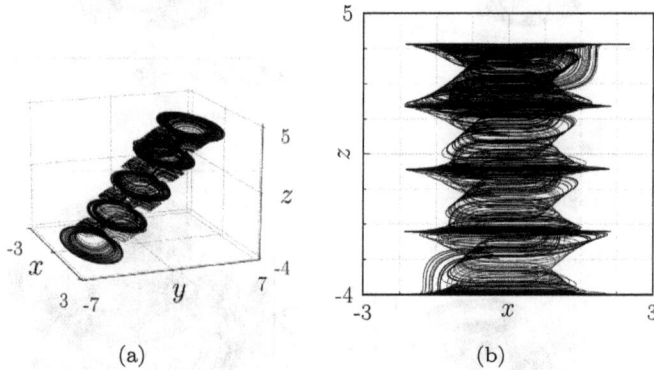

Fig. 3.17   Attractor of the system given by (3.40) and (3.47) with $a = 1, b = 10, v = 5, k = 10, m = 1, p = 2, d = 0.1$ in (a) $x - y - z$ and its projections onto the planes (b) $x - z$.

In [Escalante-González and Campos (2020b)] an approach to generate hyperchaotic attractors in higher-dimensional systems is also proposed. The approach is based on the coupling of three-dimensional systems without equilibria given by (3.40) with hidden scroll attractors. Let us consider two systems given by (3.40) $\dot{X}_1 = (\dot{x}_1, \dot{y}_1, \dot{z}_1)^T$ and $\dot{X}_2 = (\dot{x}_2, \dot{y}_2, \dot{z}_2)^T$ whose state vectors are $X_1 = (x_1, y_1, z_1)^T$ and $X_2 = (x_2, y_2, z_2)^T$, coupling both systems it is possible to generate a six dimensional system, where $X = (X_1, X_2)^T = (x_1, y_1, z_1, x_2, y_2, z_2)^T$ is the state vector of this new system.

The proposed coupling is

$$\dot{z}_1 = -x_1^2 v(z_1 - f(X_1) - k_1 g(X_2)) + d,$$
$$\dot{z}_2 = -x_2^2 v(z_2 - f(X_2) - k_2 g(X_1)) + d, \qquad (3.48)$$

where $k_1, k_2 \neq 0$ and g is a functional that joins the subsystems preserving the absence of equilibria in the new system $\dot{X}$. It is also possible to extend the approach for a larger number of coupled systems, for instance, for a nine-dimensional example, we consider the functional $g = f(X)$

and three identical subsystems $\dot{X}_1$, $\dot{X}_2$ and $\dot{X}_3$ whose state vectors are $X_1 = (x_1, y_1, z_1)^T$, $X_2 = (x_2, y_2, z_2)^T$ and $X_3 = (x_3, y_3, z_3)^T$, then one possible coupling is as follows:

$$\begin{aligned}
\dot{z}_1 &= -x_1^2 v(z_1 - f(X_1) - k_1 f(X_3)) + d, \\
\dot{z}_2 &= -x_2^2 v(z_2 - f(X_2) - k_2 f(X_1)) + d, \\
\dot{z}_3 &= -x_3^2 v(z_3 - f(X_2) - k_3 f(X_2)) + d,
\end{aligned} \tag{3.49}$$

with $k_1 = k_2 = k_3 = 0.1$ To illustrate this coupling, let us consider three identical systems given by (3.40) with the parameters $a = 1$, $b = 10$, $v = 5$, $m = 1$, $p = 2$ and $d = 0.1$ i.e. they exhibit a hidden double-scroll attractor. By choosing $k_1 = k_2 = k_3 = 0.1$ the new system exhibits a scroll attractor shown in Fig. 3.18. Note that, if all k_i are chosen identically, then the coupled systems evolve equally for some initial conditions such that $X_i(0) = X_j(0)$, $\forall i \neq j$.

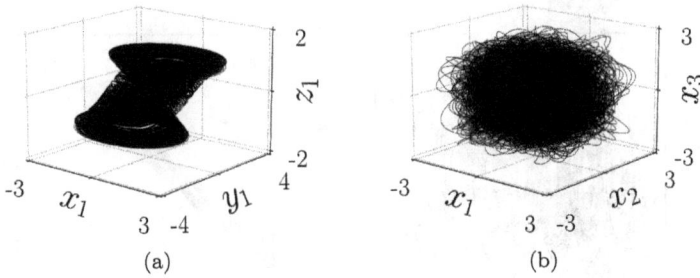

(a)   (b)

Fig. 3.18   Attractor of the system given by (3.40) with the coupling given by (3.49) and $k_1 = k_2 = k_3 = 0.1$, $a = 1$, $b = 10$, $v = 5$, $m = 1$, $p = 2$ and $d = 0.1$ in (a) $x_1 - y_1 - z_1$ and (b) $y_1 - y_2 - y_3$.

In [Zhang and Wang (2019)] it is reported a class of systems with hidden hyperchaotic attractors n a system with an infinite number of equilibria. This type of attractors can be considered "hidden" from a computational point of view [Jafari and Sprott (2013)] due to the difficulties for the attractor localization by selecting initial conditions close to the equilibria. However, this type of attractors does not fulfill the definition of hidden attractors. The system is described as follows:

$$\begin{aligned}
\dot{x} &= y, \\
\dot{y} &= z - \alpha yzw, \\
\dot{z} &= -x - y - \beta z + f(X, M, N), \\
\dot{w} &= y + z,
\end{aligned} \tag{3.50}$$

where $\alpha, \beta \in \mathbb{R}$ and $f(X, M, N)$ is a step sequence with the middle point of its vertical edge located at the origin of the coordinate axis given by

$$f(X, M, N) = 0.5\,\text{sgn}(x) + 0.5 \sum_{N}^{n=1}[\text{sgn}(x-n)+1] + 0.5 \sum_{M}^{m=1}[\text{sgn}(x+m)-1].$$

(3.51)

The equilibrium of the system $X^* = (x^*, 0, 0, c)$, where $c\mathbb{R}$ and $x^* = -M - 0.5, -M + 0.5, \ldots, -0.5, 0.5, \ldots, N - 0.5, N + 0.5$, i.e. there are lines along the w-axis direction with an infinite number of equilibria.

To illustrate this class of systems let us consider $\alpha = 0.01$, $\beta = 0.62$, $N = 2$, $M = 2$ and the initial condition $X_0 = [0.4, 0, 0, -0.4]^T$. The computed Lyapunov exponents are $\{0.0837, 0.0791, 0, -0.7728\}$. Two projections of the multi-scroll hidden attractor onto the planes $x - y$ and $x - w$ are shown in the Figs. 3.19(a) and 3.19(b), respectively.

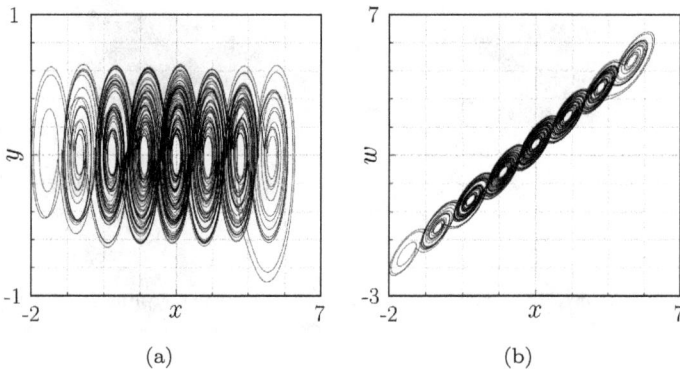

(a)   (b)

Fig. 3.19   Attractor of the system given by (3.50) and (3.51) with $\alpha = 0.01$, $\beta = 0.62$, $N = 2$, $M = 2$ and the initial condition $X_0 = [0.4, 0, 0, -0.4]^T$ projected onto the planes (a) $x - y$ and (b) $x - w$.

As with self-excited attractors, it is also possible to generate multi-scroll attractors in more than one direction. When scrolls are distributed along more than one direction the attractors are usually called grid attractor.

In [Escalante-González and Campos-Cantón (2019)] a class of systems without equilibria with hidden grid attractors was reported. This class of systems are PWL given by

$$\dot{x} = A_i x + B_i, \quad x \in P_i,$$

(3.52)

where each $A_j = A$ is given as follows:

$$A = \begin{pmatrix} \frac{a+c}{2} & -b & \frac{c-a}{2} \\ \frac{b}{2} & a & \frac{-b}{2} \\ \frac{c-a}{2} & b & \frac{a+c}{2} \end{pmatrix}, \tag{3.53}$$

where $a, b \in \mathbb{R} - \{0\}$ and $c \in \mathbb{R}$. The eigenvalues of the operator A are $\lambda_1 = c, \lambda_{2,3} = a \pm ib$. Each B_i is defined trough a function $B(x)$ as follows:

$$\begin{aligned} B(x) = &-f_1(x)a_1 - f_2(x, f_1)a_2 - f_3(x)a_3 \\ &+ f_4(x, f_1, f_3)p_1 - f_5(x, f_1, f_3)(a_1 - a_3), \end{aligned} \tag{3.54}$$

where $f_1, \dots, f_5$ are the functions defined next.

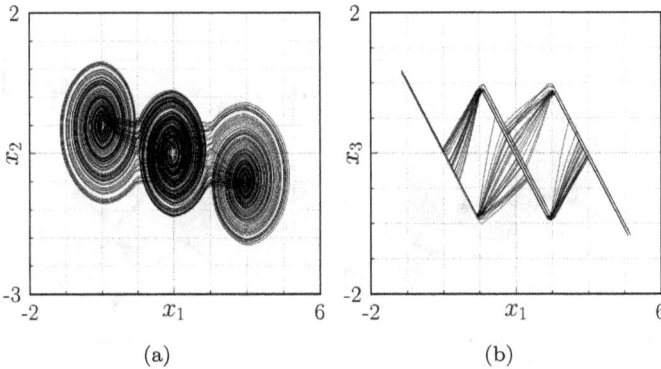

(a)   (b)

Fig. 3.20 1D hidden multi-scroll attractor given by (3.52), (3.53) and (3.54) with $f_2 = f_3 = 0$, $p = 2$, $a = 0.7$, $b = 10$, $c = 0$, $v = 11$, $w = 0.1$, $k = 0.25$ and $\Delta x_1 = 2$.

$$f_1(x) = \begin{cases} 0, & \text{if} \quad n_{f_1}x \le C_{11}; \\ \Delta_{x_1}, & \text{if } C_{11} < n_{f_1}x \le C_{12}; \\ \vdots & \\ p\Delta_{x_1}, & \text{if} \quad C_{1p} < n_{f_1}x; \end{cases} \tag{3.55}$$

where $\Delta_{x_1} \in \mathbb{R}_{>0}$, S_{1i} for $i = 1, \dots, p$ with $p \in \mathbb{R}$ are the switching surfaces defined as $S_{1i} = \{x \in \mathbb{R}^3 | x_1 = C_{1i}\}$ with $C_{1i} = \frac{(2i-1)\Delta_{x_1}}{2}$ and $n_{f_1} = [1, 0, 0]$.

$$f_2(x, f_1) = \begin{cases} -kf_1(x), & \text{if} \quad n_{f_2}x \le C_{21}; \\ \Delta_{x_2} - kf_1(x), & \text{if } C_{22} < n_{f_2}x \le C_{22}; \\ \vdots & \\ q\Delta_{x_2} - kf_1(x), & \text{if} \quad n_{f_2}x > C_{2q}; \end{cases} \tag{3.56}$$

where $\Delta_{x_2} \in \mathbb{R}_{>0}$, S_{2i} for $i = 1, \ldots, q$ with $q \in \mathbb{R}$ are the switching surfaces defined as $S_{2i} = \{x \in \mathbb{R}^3 | x_2 = C_{2i}\}$ with $C_{2i} = \frac{(2i-1)\Delta_{x_3}}{2} - kf_1(x)$, $k \in \mathbb{R}$ and $n_{f_2} = [0, 1, 0]$.

$$f_3(x) = \begin{cases} 0, & \text{if} \quad n_{f_3}x \leq S_{31}; \\ \Delta_{x_3}, & \text{if } S_{31} < n_{f_3}x \leq S_{32}; \\ \vdots \\ r\Delta_{x_3}, & \text{if} \quad n_{f_3}x > S_{3r}; \end{cases} \tag{3.57}$$

where $\Delta_{x_3} \in \mathbb{R}_{>0}$, S_{3i} for $i = 1, \ldots, r$ with $r \in \mathbb{R}$ are the switching surfaces defined as $S_{3i} = \{x \in \mathbb{R}^3 | x_3 = C_{3i}\}$ with $C_{3i} = \frac{(2i-1)\Delta_{x_3}}{2}$ and $n_{f_3} = [0, 0, 1]$.

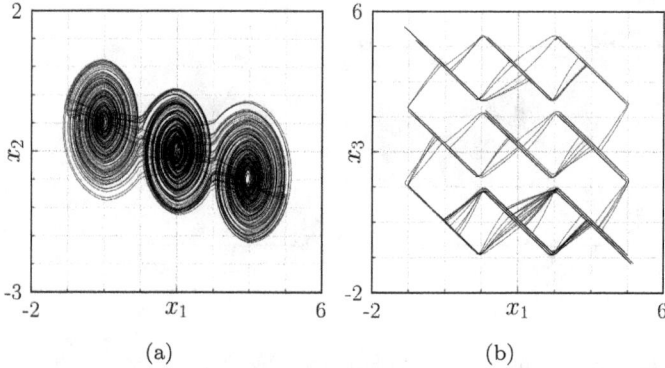

(a)   (b)

Fig. 3.21   2D hidden multi-scroll attractor given by (3.52), (3.53) and (3.54) with $f_2 = 0$, $p = r = 2$, $a = 0.7$, $b = 10$, $c = 0$, $v = 11$, $w = 0.1$, $k = 0.25$, $\Delta x_1 = 2$ and $\Delta x_3 = 2.2$.

$$f_4(x, f_1, f_3) = \begin{cases} v, & \text{if } n_{f_4}x < C_4; \\ 0, & \text{if } n_{f_4}x = C_4; \\ -v, & \text{if } n_{f_4}x > C_4; \end{cases} \tag{3.58}$$

where $v \in \mathbb{R}_{>0}$ and the switching surface $S_4 = \{x \in \mathbb{R}^3 | x_1 + x_3 = C_4\}$ with $C_4 = f_1(x) + f_3(x)$ and $n_{f_4} = [1, 0, 1]$.

$$f_5(x, f_1, f_3) = \begin{cases} -w, & \text{if } n_{f_5}x \leq C_5; \\ w, & \text{if } n_{f_5}x > C_5; \end{cases} \tag{3.59}$$

where $w \in \mathbb{R}_{>0}$ and the switching surface $S_5 = \{x \in \mathbb{R}^3 | -x_1 + x_3 = C_5\}$ with $C_5 = f_3(x) - f_1(x)$ and $n_{f_5} = [1, 0, -1]$. For 1D multi-scroll attractors

$f_2 = f_3 = 0$ or $f_1 = f_2 = 0$, for 2D multi-scroll at least a function should be equal to zero. As a first example of a 1D directional multi-scroll attractor this class of system let $p = 2$ and $q = r = 0$, $a = 0.7$, $b = 10$, $c = 0$, $v = 11$, $w = 0.1$, $k = 0.25$, $\Delta x_1 = \Delta x_2 = 2$ and $\Delta x_3 = 2.2$. The system has no equilibria and exhibits a hidden multi-scroll attractor with an arrangement of 3 scrolls along the x_1 direction. The resulting attractor is shown in Fig. 3.20. Now consider the same parameters but $p = r = 2$ and $q = 0$, thus the system exhibits a hidden multi-scroll attractor with an arrangement of 3×3 scrolls along the x_1 and x_3 directions. The resulting attractor is shown in Fig. 3.21. Finally consider $p = q = r = 2$, then a 3D hidden multi-scroll attractor is generated. The resulting attractor is shown in Fig. 3.22.

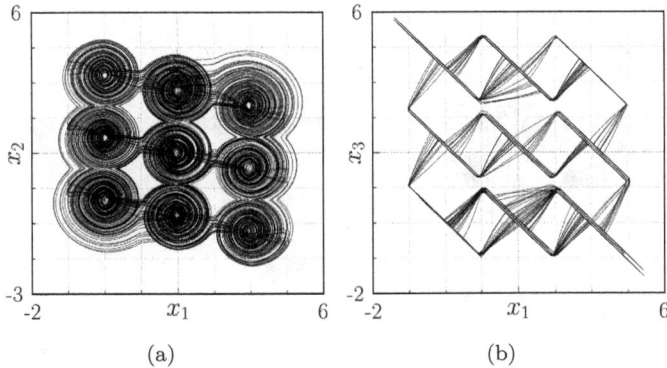

(a) (b)

Fig. 3.22 3D hidden multi-scroll attractor given by (3.52), (3.53) and (3.54) with $p = q = r = 2$, $a = 0.7$, $b = 10$, $c = 0$, $v = 11$, $w = 0.1$, $k = 0.25$, $\Delta x_1 = \Delta x_2 = 2$ and $\Delta x_3 = 2.2$.

3.3 Multi-stability in systems with hidden attractors

If for certain application it is desired the existence of more than one hidden attractor, some of the previous techniques can be used to generate more than one scroll attractor and thus producing a bi-stable or multi-stable behavior.

For instance consider the system introduced in the previous section which was based on the existence of "saddle-focus like" behavior. By modifying the system with the double scroll attractor we can obtain multiple double scroll attractors. Thus let us rewrite the system with the modifications as follows:

$$\dot{X} = \begin{cases} AX + V + W_1, & \text{if } x_3 < 0, x_1 < f(x); \\ AX + V + W_2, & \text{if } x_3 < 0, C < d + f(x), x_1 \geq f(x); \\ AX + W_1, & \text{if } x_3 = 0, x_1 < f(x); \\ AX + W_2, & \text{if } x_3 = 0, C < d + f(x), x_1 \geq f(x); \\ AX - V + W_1, & \text{if } x_3 > 0, C < d + f(x), x_1 < f(x); \\ AX - V + W_2, & \text{if } x_3 > 0, C < d + f(x), x_1 \geq f(x); \\ AX + V + W_3, & \text{if } x_3 < (d/n_3), C \geq d + f(x), x_1 < d + f(x); \\ AX + V + W_4, & \text{if } x_3 < (d/n_3), C \geq d + f(x), x_1 \geq d + f(x); \\ AX + W_3, & \text{if } x_3 = (d/n_3), C \geq d + f(x), x_1 < d + f(x); \\ AX + W_4, & \text{if } x_3 = (d/n_3), x_1 \geq d + f(x); \\ AX - V + W_3, & \text{if } x_3 > (d/n_3), C \geq d + f(x), x_1 < d + f(x); \\ AX - V + W_4, & \text{if } x_3 > (d/n_3), x_1 \geq d + f(x); \end{cases}$$

$$(3.60)$$

where $X = [x_1, x_2, x_3]^T$ is the state vector $C = n_1 x_1 + n_3 x_3$ and

$$A = \begin{pmatrix} m & -n & 0 \\ n & m & 0 \\ 0 & 0 & 0 \end{pmatrix}, \quad A = (a_1, a_2, a_3), \quad (3.61)$$

$$\begin{aligned} W_1 &= (-0.1 - f(x))a_1, \\ W_2 &= (0.1 - f(x))a_1, \\ W_3 &= (-0.1 - d - f(x))a_1, \\ W_4 &= (0.1 - d - f(x))a_1, \\ V &= [0, 0, v]^T, \end{aligned} \quad (3.62)$$

with $v, m, n, n_1, n_3 > 0$. The parameter d, n_1, n_3 determines the separation between the centers of scrolls. $f(X)$ is a functional defined as

$$f(X) = \sum_{i=1}^{M} \frac{1}{2} H\left(x_1 - \Delta_x/2 + (i-1)\Delta_x\right), \quad (3.63)$$

where $M+1$ is the number of double scroll attractors and Δ_x its separation. $H(X)$ is the step function

$$H(x) = \begin{cases} 1, & \text{if } x \geq 0, \\ -1, & \text{if } x < 0. \end{cases} \quad (3.64)$$

To illustrate this class of systems, let us consider the parameters $v = 3$, $m = 0.5$, $n = 9$, $n_1 = 1$, $n_3 = 0.5$, $d = 1$, $\Delta_x = 6$ and $M = 1$. Thus, the system present two hidden double-scroll attractors shown in the Fig. 3.23(a). Changing $M = 2$, $M = 3$ and $M = 4$ the system exhibits three, four and five hidden double-scroll attractors which are shown in the Figs. 3.23(b), 3.23(c) and 3.23(d), respectively.

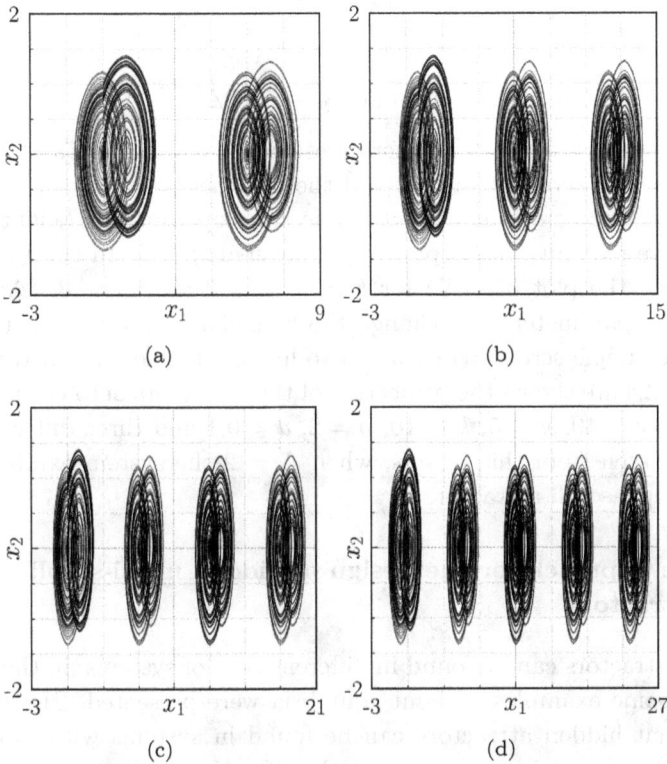

Fig. 3.23 Projection onto the plane $x_1 - x_2$ of the hidden double-scroll attractors of the system (3.60), (3.61), (3.62) and (3.63) with $v = 3$, $m = 0.5$, $n = 9$, $n_1 = 1$, $n_3 = 0.5$, $d = 1$, $\Delta_x = 6$ and (a) $M = 1$, (b) $M = 2$, (c) $M = 3$, (d) $M = 4$.

The system class of systems reported in [Escalante-González and Campos (2020b)] discussed in the previous section can also be modified in order to present more than one hidden scroll attractor. To illustrate this, let us rewrite the system as follows:

$$\dot{x} = ax + b(y - f(X)),$$
$$\dot{y} = -bx,$$
$$\dot{z} = -x^2v(z - f(X)) + d,$$
(3.65)

where $a, b, v > 0$, $b > a$. $f(X)$ is still a functional, but it defines not only the scrolls but the number of attractors, thus for instance, if two double-scroll attractors are desired a possible $f(X)$ is:

$$f(X) = \tanh\left(k(y + z - 2\Delta - \frac{p}{2}\right)$$
$$+ \tanh\left(k(y + z)\right)$$
$$+ \tanh\left(k(y + z + 2\Delta - \frac{p}{2})\right),$$
(3.66)

where the parameter $\Delta \in \mathbb{R}$ determines the separation in y and z between the two double-scroll attractors and the parameters $k, p \in \{X \in \mathbb{R} : X > 0\}$. The parameters k and p determine the slope and the height between consecutive plateaus in the plot of $f(X)$ versus $y + z$. In the Fig. 3.24(a) it is shown the plot of $f(X)$ for $k = 5$, $p = 2$ and $\Delta = 8$. Δ is also a bifurcation parameter that change the behavior of the system to exhibit one hidden triple-scroll attractor or two hidden double-scroll attractors. In the Fig. 3.24 are shown the projection of the attractors onto the plane $x - y$ for $a = 1$, $b = 10$, $v = 5$, $k = 10$, $p = 2$, $d = 0.1$ and three different values of Δ. As seen from this figures, when $\Delta = 2$ the system exhibits only a hidden triple-scroll attractor.

3.4   An approach for the design of hidden multi-scroll attractors

Hidden attractors can be found in different type of systems, in the previous sections some examples without equilibria were presented. However, it is known that hidden attractors can be found in systems with any type of equilibria and even in coexistence with self-excited attractors.

In [Ontañón-García and Campos-Cantón (2017)], the basin of attraction of self-excited scroll attractors in a class of PWL systems was studied. In this work it was found that increasing the separation between two double-scroll attractors in a bi-stable system leads to the emergence of a hidden double-scroll attractor.

Let us consider $P = \{D_1, \ldots, D_\eta\}$ such that $\mathbb{R}^3 = \mathbb{U}_k^\eta = 1$. The system studied in [Ontañón-García and Campos-Cantón (2017)] is given by

$$\dot{X} = AX + B(X), \quad X = [x, y, z]^T$$
(3.67)

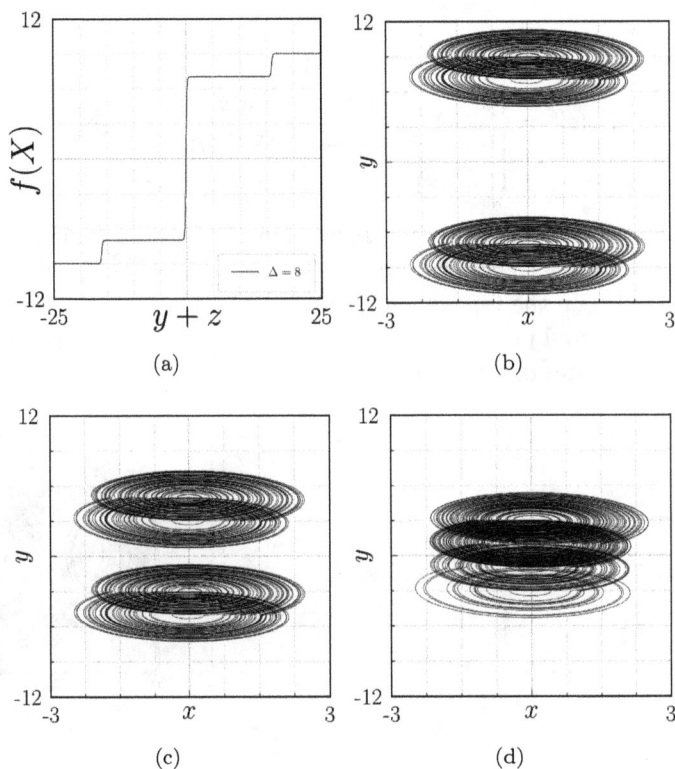

Fig. 3.24 (a) Plot of (3.66) for $k = 5$, $p = 2$ and $\Delta = 8$. Projection onto the plane $x - y$ of the hidden scroll attractors of the system (3.65) and (3.66), with $a = 1$, $b = 10$, $v = 5$, $k = 10$, $p = 2$, $d = 0.1$ and (b) $\Delta = 2$, (c) $\Delta = 2$, (d) $\Delta = 2$.

with

$$A = \begin{pmatrix} 1 & 0 & 0 \\ 0 & 1 & 0 \\ -1.5 & -1 & -1 \end{pmatrix}, \tag{3.68}$$

and

$$B(X) = \begin{pmatrix} 0 \\ 0 \\ B_c(x) \end{pmatrix} = \begin{cases} B_1, & \text{if } X \in D_1, \\ B_2, & \text{if } X \in D_2, \\ \vdots & \vdots \\ B_\eta, & \text{if } X \in D_\eta, \end{cases} \tag{3.69}$$

where

$$B_c(x) = \begin{cases} \delta + 1, & \text{if } 2/3(\delta + 1/2) \leq x, \\ \delta, & \text{if } 0 \leq < 2/3(\delta + 1/2) \leq x, \\ -\delta & \text{if } -2/3(\delta + 1/2) \leq x < 0, \\ -\delta - 1, & \text{if } -2/3(\delta + 1/2) < x. \end{cases} \quad (3.70)$$

The parameter δ determines the separation of the scrolls. For a value of $\delta = 55$ the double-scroll attractors are located around $[-37.666, 0, 0]^T$ and $[-37.666, 0, 0]^T$. In the Figs. 3.25(a) and 3.25(b) are shown the two double-scroll attractors and in the Fig. 3.25(c) are shown the three scroll attractor, two self-excited and one hidden.

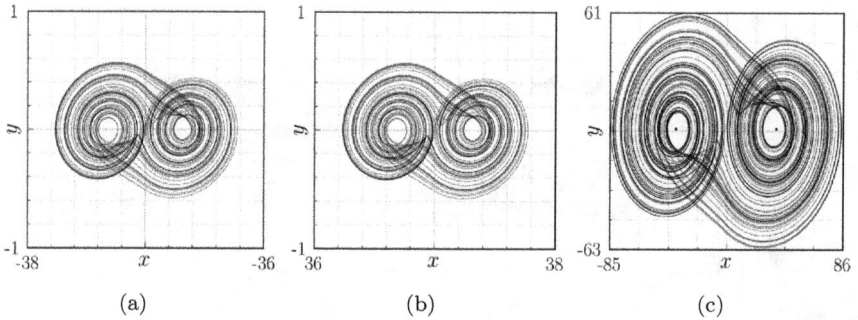

Fig. 3.25   System given by (3.67), (3.68), (3.69) and (3.70) with $\delta = 55$. (a) Self-excited double-scroll at $[-37.666, 0, 0]^T$. (b) Self-excited double-scroll at $[37.666, 0, 0]^T$. (c) are shown the three-scroll attractor, two self-excited and one hidden.

Now let us consider $X \subset \mathbb{R}^3$ and a finite partition $P = \{P_1, \ldots, P_\eta\}$ ($\eta > 1$) such that $X = \bigcup_{1 \leq i \leq \eta} P_i$, and $P_i \cap P_j = \emptyset$ for $i \neq j$. Each element of the set P is called an atom, which contains a saddle equilibrium point.

Let us consider the PWL system reported in [Escalante-González and Campos-Cantón (2019)] $T : X \to X$, whose dynamics is given by a family of subsystems as follows

$$\dot{\mathbf{x}} = A\mathbf{x} + f(\mathbf{x})B, \quad (3.71)$$

where $\mathbf{x} = (x_1, x_2, x_3)^T \in \mathbb{R}^3$ is the state vector and A is a linear operator given by

$$A = \begin{pmatrix} \frac{a}{3} + \frac{2c}{3} & b & \frac{2c}{3} - \frac{2a}{3} \\ -\frac{b}{3} & a & \frac{2b}{3} \\ \frac{c}{3} - \frac{a}{3} & -b & \frac{2a}{3} + \frac{c}{3} \end{pmatrix}, \quad (3.72)$$

with $a, b > 0$ and $c < 0$. The vector B is constant and is defined as follows

$$B = \begin{pmatrix} -\frac{a}{3} - \frac{2c}{3} \\ \frac{b}{3} \\ \frac{a}{3} - \frac{c}{3} \end{pmatrix}. \tag{3.73}$$

f is a functional given by

$$f(\mathbf{x}) = \begin{cases} -\alpha, \mathbf{x} \in P_1; \\ \alpha, \mathbf{x} \in P_2; \end{cases} \tag{3.74}$$

with $\alpha > 0$. The equilibria is given by $\mathbf{x}^*_{eq_i} = (x^*_{1_{eq_i}}, x^*_{2_{eq_i}}, x^*_{3_{eq_i}})^T = -f(\mathbf{x})A^{-1}B \in P_i$, with $i = 1, \ldots, \eta$. By considering the switching surface

$$\begin{aligned} SW &= \{x \in \mathbb{R}^3 : 2x_1 - x_3 = 0\}, \\ \{x \in \mathbb{R}^3 &: x_3 \geq 0\} \cap SW \in P_1, \\ \{x \in \mathbb{R}^3 &: x_3 < 0\} \cap SW \in P_2, \end{aligned} \tag{3.75}$$

along with the parameter values $a = 0.2, b = 5, c = -3, \alpha = 1$. The system present an heteroclinic loop and a double-scroll chaotic attractor.

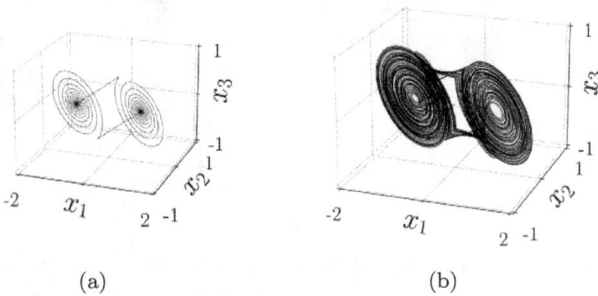

(a)   (b)

Fig. 3.26 (a) Heteroclinic loop of the system (3.71), (3.72), (3.73), (3.74) and (3.75) for the parameters $a = 0.2, b = 5, c = -3, \alpha = 1$, and in (b) the self-excited chaotic double-scroll.

Now in order to extend the system to a pair of double-scroll attractors let us consider a new functional:

$$f(\mathbf{x}) = \begin{cases} -\alpha - \gamma, \mathbf{x} \in P_1; \\ \alpha - \gamma, \quad \mathbf{x} \in P_2; \\ -\alpha + \gamma, \mathbf{x} \in P_3; \\ \alpha + \gamma, \quad \mathbf{x} \in P_4. \end{cases} \tag{3.76}$$

Now with more elements in the partition let us define the switching surfaces as:

$$SW_{12} = cl(P_1) \cap cl(P_2) = \{\mathbf{x} \in \mathbb{R}^3 : 2x_1 - x_3 = -2\gamma\},$$
$$SW_{23} = cl(P_2) \cap cl(P_3) = \{\mathbf{x} \in \mathbb{R}^3 : 2x_1 - x_3 = 0\},$$
$$SW_{34} = cl(P_3) \cap cl(P_4) = \{\mathbf{x} \in \mathbb{R}^3 : 2x_1 - x_3 = 2\gamma\}, \qquad (3.77)$$
$$SW_{i(i+1)} \cap \{\mathbf{x} \in \mathbb{R}^3 : x_3 > 0\} \in P_i,$$
$$SW_{i(i+1)} \cap \{\mathbf{x} \in \mathbb{R}^3 : x_3 \leq 0\} \in P_{i+1}.$$

Thus, with this new functional the system can present a mono-stable behavior with a self-excited multi-scroll attractor, a self-excited double-scroll attractor or bi-stability with two self-exited double-scroll attractors depending on the parameter γ. As we are interested in the generation of a hidden double-scroll attractor the bi-stable behavior is the selected one. In the Fig. 3.27 are shown the two double-scroll attractors for the parameter for $a = 0.2$, $b = 5$, $c = -3$, $\alpha = 1$ and $\gamma = 3$.

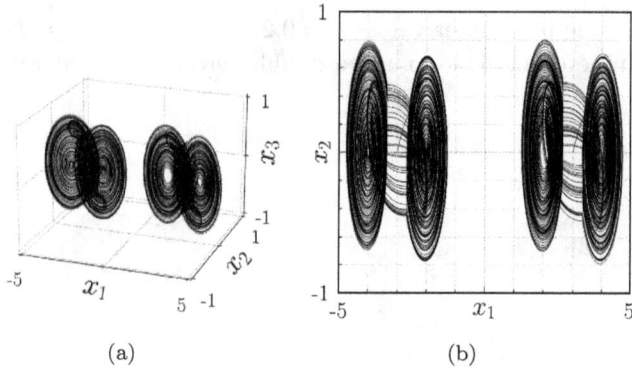

(a)   (b)

Fig. 3.27   Double-scroll attractors of the system (3.71), (3.72), (3.73), (3.76) and (3.77) for the parameters $a = 0.2$, $b = 5$, $c = -3$, $\alpha = 1$ and $\gamma = 3$ in (a) $x_1 - x_2 - x_3$ and its projections onto the plane (b) $x_1 - x_2$.

From the observations in [Ontañón-García and Campos-Cantón (2017)] it could be thought that for a certain value of δ a hidden attractor may emerge, however it is not the case. As experiment, let us take the value of $\delta = 30$ and simulate for some initial conditions. In the Fig. 3.28(a) it is shown the simulation for $\mathbf{x} = [0, 0, 0]^T$ $t \in [0, 200]$, and as it can be seen, after a transient oscillation that resembles a double-scroll the solution converges to the left self-excited attractors around $t = 180$. In the Fig. 3.28(b) it is shown the simulation for $\mathbf{x} = [0, 5, 0]^T$ $t \in [1000, 1200]$, and as it can be seen, after a transient oscillation that resembles a double-scroll the solution converges to the left self-excited attractors around $t = 1100$. Thus, depending on the initial condition the transient last more time, however the hidden attractor does not emerge. These test can be performed with larger values of δ but the result is the same.

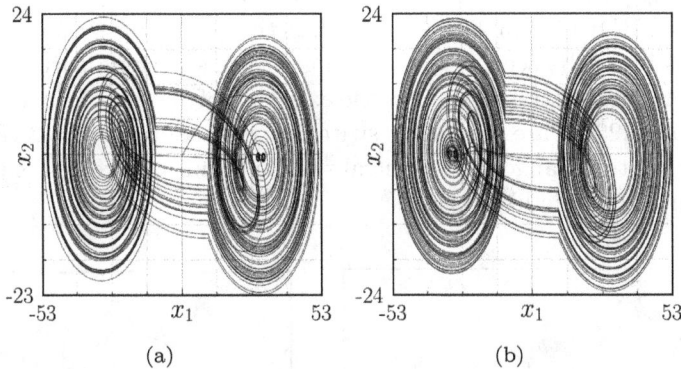

Fig. 3.28   Simulation of the system (3.71), (3.72), (3.73), (3.76) and (3.77) for the parameters $a = 0.2$, $b = 5$, $c = -3$, $\alpha = 1$ and $\gamma = 30$ for (a) $\mathbf{x} = [0, 0, 0]^T$, $t \in [0, 200]$ and (b) $\mathbf{x} = [0, 5, 0]^T$ $t \in [1000, 1200]$.

This situation was previously studied in [Escalante-González and Campos-Cantón (2019)], the study revealed that the emergence of the hidden attractor is related to the existence of trajectories that joints the neighborhood of both double-scroll attractors which seen in a larger scale resemble heteroclinic orbits. Thus, in order to allow the emergence of the hidden attractor it is required to eliminate those trajectories. For each hidden double-scroll attractor exist a heteroclinic loop, their existence depends in part to the orientation and location of the switching surface, i.e. if the orientation of the switching surface is changed the heteroclinic loops are broken. Since the switching surface between the two self-excited attractors is similar to those in the middle of the double-scrolls, a similar geometry is expected when the double-scroll attractors are seen as "equilibria" and therefore "heteroclinic like" orbits are expected. To avoid this orbits, one approach is to change the switching surface between the double-scroll attractors. Let us redefine the switching surfaces as

$$SW_{12} = cl(P_1) \cup cl(P_2) = \{\mathbf{x} \in \mathbb{R}^3 : 2x_1 - x_3 = -2\gamma, x_1 < 0\},$$
$$SW_{23} = cl(P_2) \cup cl(P_3) = \{\mathbf{x} \in \mathbb{R}^3 : x_1 = 0\},$$
$$SW_{34} = cl(P_3) \cup cl(P_4) = \{\mathbf{x} \in \mathbb{R}^3 : 2x_1 - x_3 = 2\gamma, x_1 > 0\}, \qquad (3.78)$$
$$SW_{i(i+1)} \cap \{\mathbf{x} \in \mathbb{R}^3 : x_3 > 0\} \in P_i,$$
$$SW_{i(i+1)} \cap \{\mathbf{x} \in \mathbb{R}^3 : x_3 \leq 0\} \in P_{i+1}.$$

The Fig. 3.29 illustrate the change in the definition of the switching surfaces, in the Fig. (3.29(a)) are shown the switching surfaces given by (3.77), as it can be seen, the orientation of the switching surface between the double-scroll attractors favors the existence of "heteroclinic like" orbits. In the Fig. 3.29(b) are shown the switching surfaces given by (3.78), it can be seen that the change in the central switching surface avoids the existence of the "hetero-clinic like" orbits.

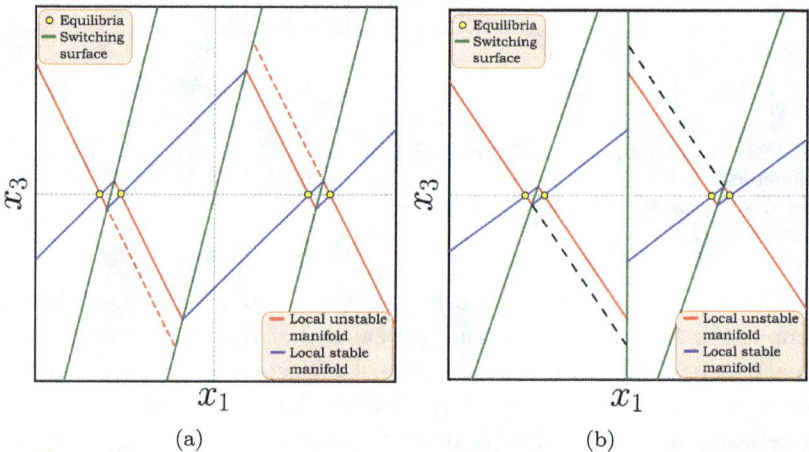

Fig. 3.29   Illustration of the system given by (3.71), (3.72), (3.73) and (3.76) with the switching given by (a) (3.77) and (b) (3.78).

In [Escalante-González and Campos (2020c)] this class of systems is generalized in order to generate not only double-scroll attractors but a hidden multi-scroll attractor inj coexistence with multiple self-excited double-scroll attractors.

Consider the system $T : X \to X$ described by

$$\dot{\mathbf{x}} = A\mathbf{x} - AB(\mathbf{x}), \qquad (3.79)$$

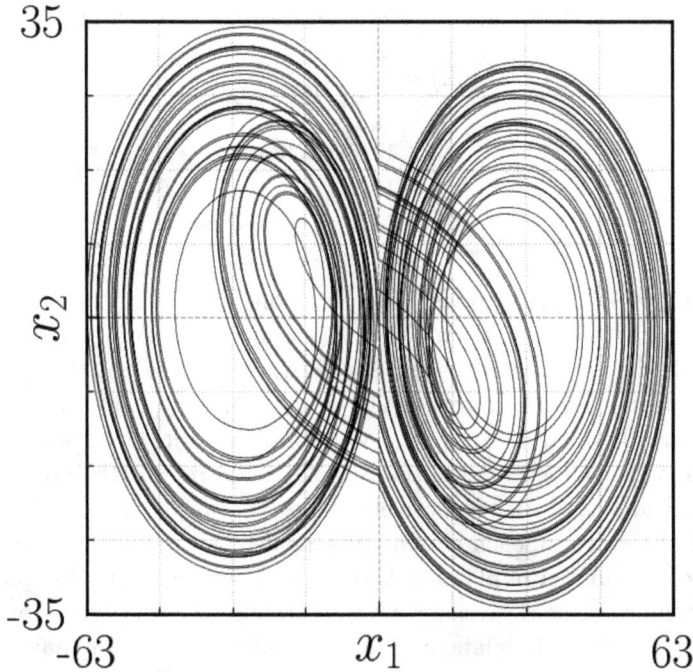

Fig. 3.30   Hidden double-scroll attractors of the system (3.71), (3.72), (3.73), (3.76) and (3.78) for the parameters $a = 0.2$,$b = 5$, $c = -3$, $\alpha = 1$ and $\gamma = 30$ with $\mathbf{x} = [0,0,0]^T$ $t \in [10000, 10100]$ projected onto the plane $x_1 - x_2$.

where the state vector is $\mathbf{x} = (x_1, x_2, x_3)^T \in \mathbb{R}^3$ and the same linear operator used previously:

$$A = \begin{pmatrix} \frac{a}{3} + \frac{2c}{3} & b & \frac{2c}{3} - \frac{2a}{3} \\ -\frac{b}{3} & a & \frac{2b}{3} \\ \frac{c}{3} - \frac{a}{3} & -b & \frac{2a}{3} + \frac{c}{3} \end{pmatrix}. \qquad (3.80)$$

In each P_i $AB(x)$ is a constant vector. The equilibria is given by $\mathbf{x}^*_{eq_i} = (x^*_{1_{eq_i}}, x^*_{2_{eq_i}}, x^*_{3_{eq_i}})^T = -B(\mathbf{x})$, with $i = 1, \ldots, \eta$. $B(\mathbf{x})$ is as follows:

$$B(\mathbf{x}) = \begin{pmatrix} B_1 \\ B_2 \\ B_3 \end{pmatrix} = \begin{pmatrix} \alpha g(2(x_1 - f_1(x_1)) - x_3, x_3) + f_1(x_1) \\ 0 \\ 0 \end{pmatrix}, \qquad (3.81)$$

where

$$f_1(x_1) = \sum_{j=1}^{N_{x_1}} \gamma u(x_1 + 2\gamma(j - 1) - \gamma(N_{x_1} - 1)). \qquad (3.82)$$

$\alpha, \gamma > 0$ and $\gamma/\alpha \geq 10$. $u(y)$ and $g(y, z)$ are step function given by:

$$u(y) = \begin{cases} 1, & \text{if } y \geq 0; \\ -1, & \text{if } y < 0; \end{cases} \tag{3.83}$$

$$g(y, z) = \begin{cases} 1, & \text{if } y > 0 \text{ and } z \geq 0; \\ -1, & \text{if } y \leq 0 \text{ and } z \geq 0; \\ 1, & \text{if } y \geq 0 \text{ and } z < 0; \\ -1, & \text{if } y < 0 \text{ and } z < 0. \end{cases} \tag{3.84}$$

The function $B(\mathbf{x})$ involves three functions, however it can be analyzed by parts to better understand what is its effect in the system. Thus, let us take B_1 the only non-zero element of $B(\mathbf{x})$ and divide it the analysis by summing terms, i.e. the term $\alpha g(2(x_1 - f_1(x_1)) - x_3, x_3)$ and $f_1(x_1)$.

To start the analysis of $f_1(x_1)$ we can observe its plots with different values of N_{x_1}, for instance, in the Figs. 3.31(a) and 3.31(b) are shown for $N_{x_1} = 3$ and $N_{x_1} = 6$, respectively. The plots resemble a stair which is centered at the origin, each plateau has the height and width which is equal to 2γ and the number of plateaus is $N_{x_1} + 1$. Therefore, $f_1(x_1)$ is generating N_{x_1} switching surfaces given by $\{\mathbf{x} \in \mathbb{R}^3, \epsilon \in \mathbb{R} : x_1 = \epsilon\}$ which are planes that are parallel to the plane $x_2 - x_3$. Thus, it can be said that the effect of the term $f_1(x)$ in $B(\mathbf{x})$ is to generate a partition $R = \{R_1, \ldots, R_{N_{x_1}+1}\}$ of X.

Now, in order to analyze the second term in B_1 let us remember that $g(y, z)$ is a step function and thus, $\alpha g(2(x_1 - f_1(x_1)) - x_3, x_3)$ generates

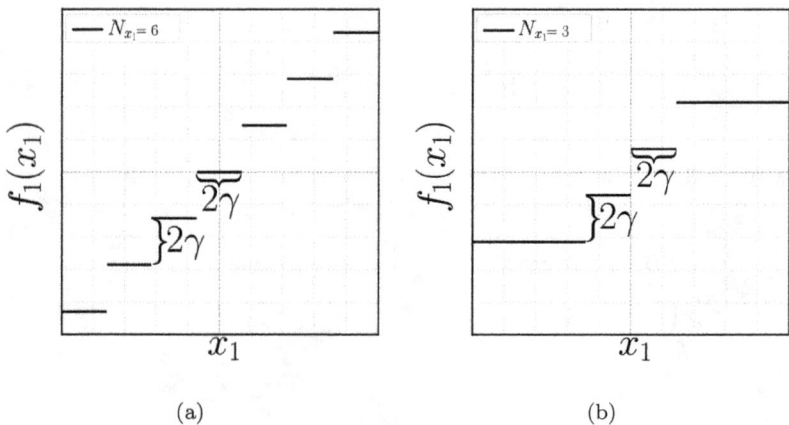

(a)   (b)

Fig. 3.31   Plot of the function (3.82) with (3.83), (3.84) for (a) $N_{x_1} = 3$ and (b) $N_{x_1} = 6$.

switching surfaces given by $\{\mathbf{x} \in R_i : 2(x_1 - f_1(x_1)) - x_3 = 0\}$ for $i = 1, \ldots, N_x + 1$. The number of switching surfaces depends on $f_1(x_1)$, since it takes $N_{x_1} + 1$ and thus the number of switching surfaces generated by $\alpha g(2(x_1 - f_1(x_1)) - x_3, x_3)$ is also $N_{x_1} + 1$, i.e. it generates a switching surface in each element of the partition R made by $f_1(x_1)$. Therefore, a new partition $P = \{P_1, \ldots, P_{2N_{x_1}+2}\}$ is generated.

Then, considering the effect of both terms, $f_1(x_1)$ and $\alpha g(2(x_1 - f_1(x_1)) - x_3, x_3)$ it can be seen that $B(\mathbf{x})$ locates the equilibria of the system along x_1, a pair of equilibrium point with a separation of 2α in the middle of each element R_i for $i = 1, \ldots, N_{x_1} + 1$. Thus, $B(\mathbf{x})$ locates an equilibrium point in each P_i for $i = 1, \ldots, 2N_{x_1} + 2$.

The equilibria is located as

$$
\begin{aligned}
x^*_{1eq_1} &= (0)2\gamma - \gamma(N_{x_1}) - \alpha, & x^*_{1eq_2} &= (0)2\gamma - \gamma(N_{x_1}) + \alpha, \\
x^*_{1eq_3} &= (1)2\gamma - \gamma(N_{x_1}) - \alpha, & x^*_{1eq_4} &= (1)2\gamma - \gamma(N_{x_1}) + \alpha, \\
x^*_{1eq_5} &= (2)2\gamma - \gamma(N_{x_1}) - \alpha, & x^*_{1eq_6} &= (2)2\gamma - \gamma(N_{x_1}) + \alpha,
\end{aligned}
$$

$$\vdots \qquad\qquad\qquad \vdots$$

$$
\begin{aligned}
x^*_{1eq_{2N_{x_1}-1}} &= (N_{x_1} - 1)2\gamma - \gamma(N_{x_1}) - \alpha, & x^*_{1eq_{2N_{x_1}}} &= (N_{x_1} - 1)2\gamma - \gamma(N_{x_1}) + \alpha, \\
x^*_{1eq_{2N_{x_1}+1}} &= (N_{x_1})2\gamma - \gamma(N_{x_1}) - \alpha, & x^*_{1eq_{2N_{x_1}+2}} &= (N_{x_1})2\gamma - \gamma(N_{x_1}) + \alpha,
\end{aligned}
$$

$$\tag{3.85}$$

while the switching surfaces are

$$
\begin{aligned}
SW_{1,2} &= \{\mathbf{x} \in \mathbb{R}^3 : 2(x_1 - (x^*_{1eq_1} + \alpha)) - x_3 = 0, x_1 < x^*_{1eq_2} + (\gamma - \alpha)\}, \\
SW_{3,4} &= \{\mathbf{x} \in \mathbb{R}^3 : 2(x_1 - (x^*_{1eq_3} + \alpha)) - x_3 \\
&= 0, x^*_{1eq_3} - (\gamma - \alpha) \le x_1 < x^*_{1eq_4} + (\gamma - \alpha)\},
\end{aligned}
$$

$$\vdots$$

$$
\begin{aligned}
SW_{2N_{x_1}+1, 2N_{x_1}+2} &= \{\mathbf{x} \in \mathbb{R}^3 : 2(x_1 - (x^*_{1eq_{N_{x_1}+1}} + \alpha)) - x_3 \\
&= 0, x^*_{eq_{2N_{x_1}+1}} - (\gamma - \alpha) \le x_1\},
\end{aligned}
$$

$$\tag{3.86}$$

and

$$
\begin{aligned}
SW_{2,3} &= \{\mathbf{x} \in \mathbb{R}^3 : x = x^*_{1eq_2} + (\gamma - \alpha)\}, \\
SW_{4,5} &= \{\mathbf{x} \in \mathbb{R}^3 : x = x^*_{1eq_4} + (\gamma - \alpha)\},
\end{aligned}
$$

$$\vdots$$

$$\tag{3.87}$$

$$
\begin{aligned}
SW_{2N_{x_1}-2, 2N_{x_1}-1} &= \{\mathbf{x} \in \mathbb{R}^3 : x = x^*_{1eq_{2N_{x_1}-2}} + (\gamma - \alpha)\}, \\
SW_{2N_{x_1}, 2N_{x_1}+1} &= \{\mathbf{x} \in \mathbb{R}^3 : x = x^*_{1eq_{2N_{x_1}}} + (\gamma - \alpha)\}.
\end{aligned}
$$

This distribution of equilibria by pairs is required to generate double-scroll attractors, where α would determine the size of the attractors. Since each pair of equilibria is separated from another pair of equilibria by a distance δ, if $\delta = \alpha$ then a multi-scroll attractor is generated. However, with sufficient separation a hidden scroll attractor emerges, the number of scrolls in the hidden attractor depends on the number of self-excited attractors.

In order to illustrate the system, let us construct a system with two self-excited attractors. Consider $a = 0.2$, $b = 5$, $c = -3$, $N_{x_1} = 1$, $\alpha = 1$ and $\gamma = 10$.

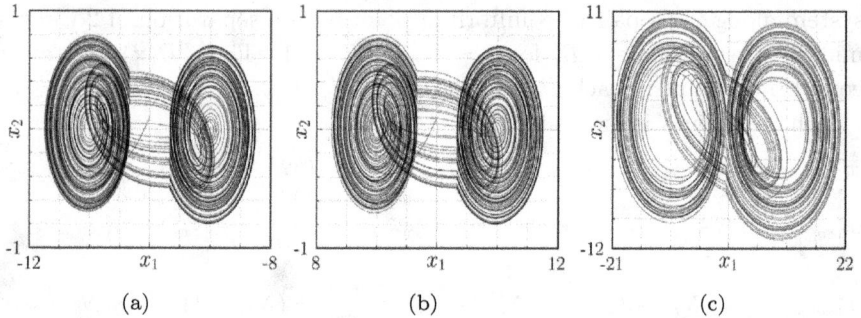

Fig. 3.32   System given by (3.79), (3.80), (3.81) and (3.82) with $a = 0.2$, $b = 5$, $c = -3$, $N_{x_1} = 1$, $\alpha = 1$ and $\gamma = 10$. (a) Self-excited double-scroll at $[-10, 0, 0]^T$ with $X_0 = [-10, 0.1, 0]^T$ and $t \in [0, 200]$. (b) Self-excited double-scroll at $[10, 0, 0]^T$ with $X_0 = [10, 0.1, 0]^T$ and $t \in [0, 200]$. (c) Hidden double-scroll attractor with $X_0 = [0, 0, 0]^T$ and $t \in [10000, 10100]$.

To see how the parameter δ allow us to modify the size of the hidden attractor without changing the self-excited attractors let us consider now $\gamma = 200$, whose plot is shown in the Fig. 3.33. Similarly, α allows us to modify the size of the self-excited attractors.

As additional examples let us consider now a higher number of scrolls in the hidden attractor, so let us consider the same parameters $a = 0.2$, $b = 5$, $c = -7$, $N_{x_1} = 1$, $\alpha = 1$ and $\gamma = 10$ with $N_{x_1} = 2$, $N_{x_1} = 3$ $N_{x_1} = 4$ and $N_{x_1} = 5$ whose attractors are plotted in Figs. 3.34(a), 3.34(b), 3.34(c) and 3.34(d), respectively.

3.5   An approach for the design of 2D and 3D directional multi-scroll attractors

In [Escalante-González and Campos (2020c)] the previous approach has been extended for the generation of 2D and 3D directional hidden multi-scroll attractors.

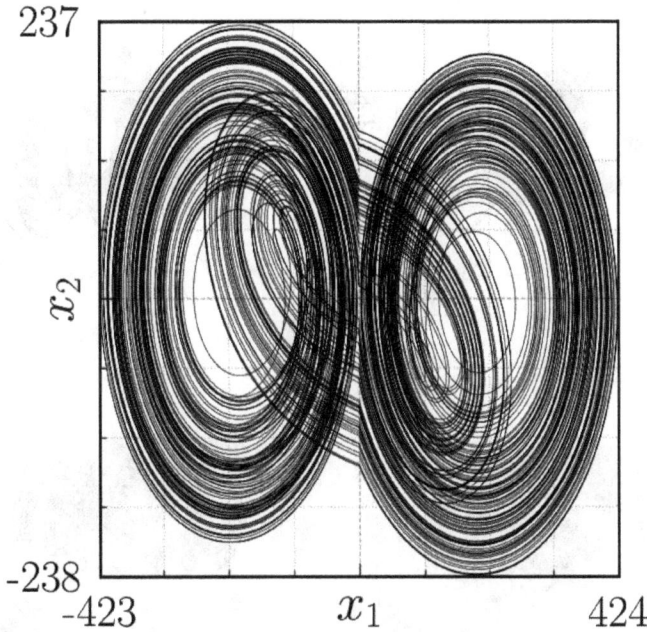

Fig. 3.33   Hidden double-scroll attractors of the system given by (3.79), (3.80), (3.81) and (3.82) with $a = 0.2$, $b = 5$, $c = -3$, $N_{x_1} = 1$, $\alpha = 1$ and $\gamma = 200$ for $X_0 = [-10, 0.1, 0]^T$ and $t \in [10000, 10300]$ projected onto the plane $x_1 - x_2$.

Consider the system $T : X \to X$ given by

$$\dot{\mathbf{x}} = A\mathbf{x} - AB(\mathbf{x}), \tag{3.88}$$

where the state vector is $\mathbf{x} = (x_1, x_2, x_3)^T \in \mathbb{R}^3$ with the linear operator A

$$A = \begin{pmatrix} \frac{a}{3} + \frac{2c}{3} & b & \frac{2c}{3} - \frac{2a}{3} \\ -\frac{b}{3} & a & \frac{2b}{3} \\ \frac{c}{3} - \frac{a}{3} & -b & \frac{2a}{3} + \frac{c}{3} \end{pmatrix}. \tag{3.89}$$

In each P_i $B(x)$ is a constant vector. The equilibria is given by $\mathbf{x}^*_{eq_i} = (x^*_{1_{eq_i}}, x^*_{2_{eq_i}}, x^*_{3_{eq_i}})^T = -B(\mathbf{x})$, with $i = 1, \ldots, \eta$. $B(\mathbf{x})$ is as follows:

$$B(\mathbf{x}) = \begin{pmatrix} f_4 + f_1 - \frac{w f_2}{\gamma} \\ f_2 \\ f_3 + \frac{w f_2}{\gamma} \end{pmatrix}, \tag{3.90}$$

with

$$f_2 = E_2 \left(\sum_{k=1}^{N_{x_2}} \gamma u(x_2 + 2\gamma(k - 1) - \gamma(N_{x_2} - 1)) \right), \tag{3.91}$$

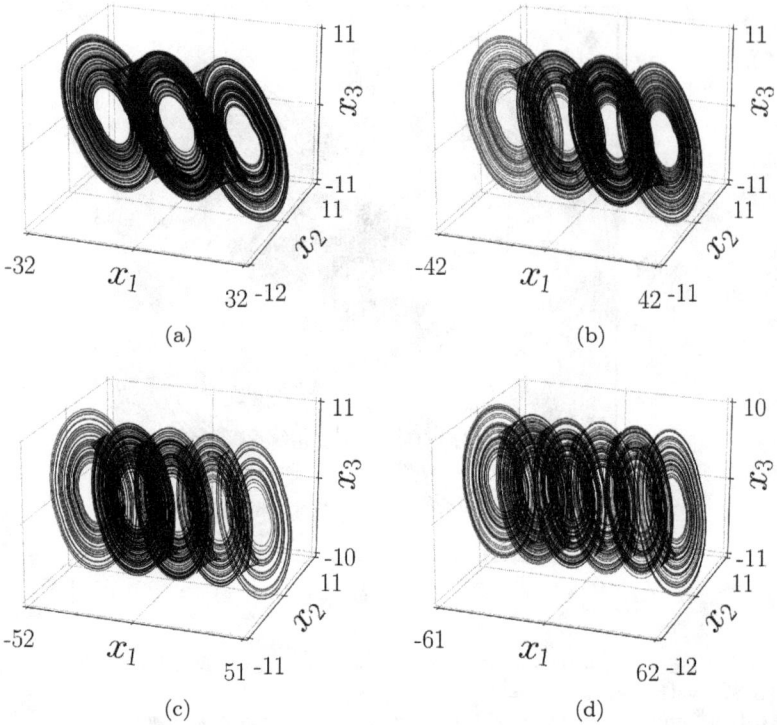

(a)

(b)

(c)

(d)

Fig. 3.34   Hidden multi-scroll attractors of the system given by (3.79), (3.80), (3.81) and (3.82) with $a = 0.2$, $b = 5$, $c = -3$, $\alpha = 1$, $\gamma = 200$ and (a) $N_{x_1} = 2$, (b) $N_{x_1} = 3$, (c) $N_{x_1} = 4$ and (d) $N_{x_1} = 5$ for $t \in [500, 1000]$ in $x_1 - x_2 - x_3$.

$$f_3 = E_3 \left(\sum_{l=1}^{N_{x_3}} \gamma u \left(\left(x_3 - \frac{w f_2}{\gamma} \right) + 2\gamma(l-1) - \gamma(N_{x_3} - 1) \right) \right), \quad (3.92)$$

$$f_1 = \sum_{j=1}^{N_{x_1}} \gamma u \left(\left(x_1 + \frac{w f_2}{\gamma} \right) + 2\gamma(j-1) - \gamma(N_{x_1} - 1) \right), \quad (3.93)$$

$$f_4 = \alpha g \left(2 \left(x_1 + \frac{w f_2}{\gamma} - f_1 \right) - \left(x_3 - \frac{w f_2}{\gamma} - f_3 \right), \left(x_3 - \frac{w f_2}{\gamma} - f_3 \right) \right), \tag{3.94}$$

where $\alpha, \gamma > 0$ and $\gamma/\alpha \geq 10$, $w \geq 0$ and $E_2, E_3 \in \{0, 1\}$. The functions $u(y)$ and $g(y, z)$ are the step functions introduced before:

$$u(y) = \begin{cases} 1, & \text{if } y \geq 0; \\ -1, & \text{if } y < 0; \end{cases} \tag{3.95}$$

$$g(y,z) = \begin{cases} 1, & \text{if } y > 0 \text{ and } z \geq 0; \\ -1, & \text{if } y \leq 0 \text{ and } z \geq 0; \\ 1, & \text{if } y \geq 0 \text{ and } z < 0; \\ -1, & \text{if } y < 0 \text{ and } z < 0. \end{cases} \qquad (3.96)$$

With this generalized approach it is possible to generate 1D, 2D and 3D scroll attractors. The parameters select in which directions the grid attractor extend E_2 and E_3 according to the following

$$\begin{aligned} E_1 &= E_2 = 0 & \text{scrolls along } x_1 \text{ direction (1D)}, \\ E_1 &= 1, E_2 = 0 & \text{scrolls along } x_1 \text{ and } x_2 \text{ directions (2D)}, \qquad (3.97) \\ E_1 &= E_2 = 1 & \text{scrolls along } x_1, /x_2 \text{ and } x_3 \text{ directions (3D)}. \end{aligned}$$

The number of scrolls in the hidden grid attractor is given by $(N_{x_1} + 1) \times (E_2 N_{x_2} + 1) \times (E_3 N_{x_3} + 1)$.

In order to describe the equilibria of the system we will denote the equilibria along the x_1 axis found before as

$$\mathbf{x}_{eq_j}^{*1D} = (x_{1eq_j}^{*1D}, x_{2eq_j}^{*1D}, x_{3eq_j}^{*1D})^T, \quad j = 1 \dots, 2N_{x_1} + 2 \qquad (3.98)$$

$\mathbf{x}_{eq_j}^{*1D} = (x_{1eq_j}^{*1D}, x_{2eq_j}^{*1D}, x_{3eq_j}^{*1D})^T$ for $j = 1 \dots, 2N_{x_1} + 2$. With this notation we can write the rest of the equilibria as

$$\mathbf{x}_{jkl}^{*3D} = \mathbf{x}_{eq_j}^{*1D} + \begin{pmatrix} 0 \\ 0 \\ (E_3 l)2\gamma - \gamma N_{x_3} \end{pmatrix} + \begin{pmatrix} 0 \\ (E_2 k)2\gamma - \gamma N_{x_2} \\ 0 \end{pmatrix} + \begin{pmatrix} -\frac{w f_2(y)}{\gamma} \\ 0 \\ \frac{w f_2(y)}{\gamma} \end{pmatrix},$$

$$(3.99)$$

where $k = 1, \ldots, 2N_{x_2} + 2$ and $l = 1, \ldots, 2N_{x_3} + 2$.

As an example for a 2D hidden grid attractor with scrolls along the x_1 and x_2 directions let us consider $a = 0.2$, $b = 5$, $c = -7$, $w = 0.15$ with $E_2 = 1$, $E_3 = 0$, $N_{x_1} = 2$ and $N_{x_3} = 2$. The hidden attractor is shown in the Fig. 3.35(a). By replacing $E_2 = 0$, $E_3 = 1$ the scrolls extend along the x_1 and x_3 directions as shown in the Fig. 3.35(b).

Now to illustrate the construction of 3D hidden grid attractor, let us consider $E_2 = 1$, $E_3 = 1$. The resulting hidden attractor is shown in the Fig. 3.36.

3.6 An approach for the design of Nested hidden attractors

In this section an approach for the design of Nested hidden attractors reported in [Escalante-González and Campos (2021)] will be addressed. The approach is based on PWL systems, thus let us define first a finite partition

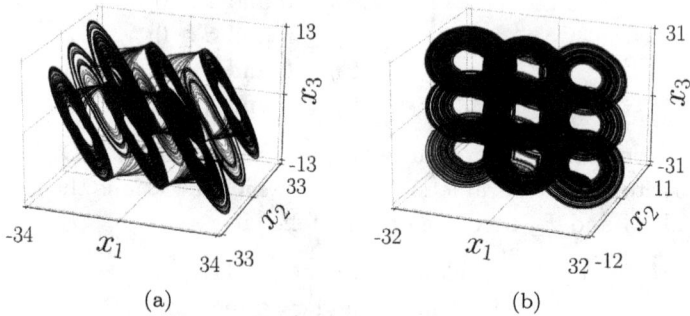

Fig. 3.35 2D hidden grid attractor of the system given by (3.88), (3.89) and (3.90) with $a = 0.2$, $b = 5$, $c = -7$, $w = 0.15$, $N_{x_1} = 2$ and $N_{x_3} = 2$ with (a) $E_2 = 1$, $E_3 = 0$ and (b) $E_2 = 0$, $E_3 = 1$.

of $X \subset \mathbb{R}^3$, $P = \{P_1, \ldots, P_\eta\}$, with $\eta > 1$ and $\eta \in \mathbb{N}$. The class of PWL system $T : X \to X$ used for the approach can be described by

$$\dot{\mathbf{x}} = A\mathbf{x} + f(\mathbf{x})B, \tag{3.100}$$

where the state vector is $\mathbf{x} = (x_1, x_2, x_3)^T \in \mathbb{R}^3$. The linear operator $A = \{\alpha_{ij}\} \in \mathbb{R}^{3\times 3}$ is given by

$$A = \begin{pmatrix} \frac{a}{3} + \frac{2c}{3} & b & \frac{2c}{3} - \frac{2a}{3} \\ -\frac{b}{3} & a & \frac{2b}{3} \\ \frac{c}{3} - \frac{a}{3} & -b & \frac{2a}{3} + \frac{c}{3} \end{pmatrix}, \tag{3.101}$$

where $a, b > 0$ and $c < 0$, such that the complexification of A, $A^{\mathbb{C}}$ has the eigenvalues $\lambda_1 = a + ib$, $\lambda_2 = a - ib$ and $\lambda_3 = c$. The vector B is constant and is given by

$$B = \begin{pmatrix} -\frac{a}{3} - \frac{2c}{3} \\ \frac{b}{3} \\ \frac{a}{3} - \frac{c}{3} \end{pmatrix}, \tag{3.102}$$

$f(\mathbf{x})$ is a functional and is given by:

$$f(\mathbf{x}) = \alpha u \left(2(x_1 - \sum_{i=1}^{m} w_i) - x_3, \ x_3 \right) + \sum_{i=1}^{m} w_i, \tag{3.103}$$

with $\alpha \in \mathbb{R}$, $m \in \mathbb{N}$, and w_i are as follows

$$w_0 = 0, \tag{3.104}$$

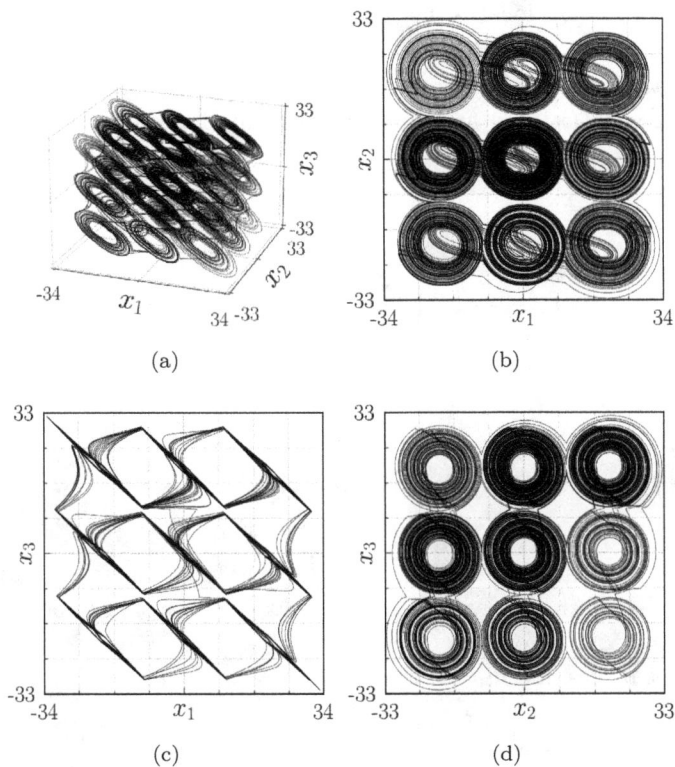

Fig. 3.36   3D hidden grid attractor of the system given by (3.88), (3.89) and (3.90) with $a = 0.2$, $b = 5$, $c = -7$, $w = 0.15$, $N_{x_1} = 2$, $N_{x_3} = 2$, $E_2 = 1$, $E_3 = 1$ in (a) $x_1 - x_2 - x_3$ and its projections onto the planes (b) $x_1 - x_2$, (c) $x_1 - x_3$ and (d) $x_2 - x_3$.

$$w_i = \gamma^{m+1-i} u \left(x_1 - \sum_{j=0}^{i-1} w_j, x_3 \right), \quad \text{for } i = 1, \ldots, m, \text{ and } m \geq 1,$$

$$(3.105)$$

with $\gamma \in \mathbb{R}$, $u(x_1, x_3)$ is a step function that returns 1 or -1 and is defined as follows:

$$u(x, z) = \begin{cases} 1, & \text{if } x > 0 \text{ and } z \geq 0; \\ -1, & \text{if } x \leq 0 \text{ and } z \geq 0; \\ 1, & \text{if } x \geq 0 \text{ and } z < 0; \\ -1, & \text{if } x < 0 \text{ and } z < 0. \end{cases} \qquad (3.106)$$

The functional $f(\mathbf{x})$ locates the saddle-foci as $\mathbf{x}_{eq_i}^* = (x_{1_{eq_i}}^*, x_{2_{eq_i}}^*, x_{3_{eq_i}}^*)^T = -f(\mathbf{x})A^{-1}B \in P_i$, with $i = 1, \ldots, \eta$. The approach is some pairs of self-excited attractors that serves as base for the emergence of hidden attractors and therefore, $f(\mathbf{x})$ has an important role in the approach. In order to understand how $f(\mathbf{x})$ works let us rewrite it as

$$f(\mathbf{x}) = f_1 + f_2, \tag{3.107}$$

where

$$f_1 = \alpha u \left(2 \left(x_1 - \sum_{i=1}^{m} w_i \right) - x_3, \ x_3 \right), \tag{3.108}$$

and

$$f_2 = \sum_{i=1}^{m} w_i. \tag{3.109}$$

Consider f_2 first, and let us assume m_1, then

$$f_2 = w_1 = \gamma u(x_1, x_3), \tag{3.110}$$

since $u(x_1, x_3)$ is a step function, it can be seen that it generates a partition $G = \{G_1, G_2\}$ with the switching surface $\{\mathbf{x} \in \mathbb{R}^3 : x_1 = 0\}$ such that for $\mathbf{x}$ in G_1 $f_2 = -\gamma$, and for $\mathbf{x}$ in G_2, $f_2 = \gamma$.

Now let us assume $m = 2$, then

$$w_1 = \gamma^2 u(x_1, x_3), \tag{3.111}$$

$$w_2 = \gamma u(x_1 - \gamma^2 u(x_1, x_3), x_3), \tag{3.112}$$

$$f_2 = \gamma u(x_1 - \gamma^2 u(x_1, x_3), x_3) + \gamma^2 u(x_1, x_3). \tag{3.113}$$

We can analyze this f_2 by each summing term. Consider first the term w_1, it is somehow similar to the previous case with $m = 1$, it generates a partition $G = \{G_1, G_2\}$ with the switching surface $\{\mathbf{x} \in \mathbb{R}^3 : x_1 = 0\}$ such that for $\mathbf{x}$ in G_1 $f_2 = -\gamma^2$, and for $\mathbf{x}$ in G_2, $f_2 = \gamma^2$. Now, consider term w_2, it generates a partition in G_1 as $G_1 = \{G_{11}, G_{12}\}$ with the switching surface $\{\mathbf{x} \in \mathbb{R}^3 : x_1 = w_1 = -\gamma^2\}$, but also generates a partition in G_2 as $G_2 = \{G_{21}, G_{22}\}$ with the switching surface $\{\mathbf{x} \in \mathbb{R}^3 : x_1 = w_1 = \gamma^2\}$. Therefore, the location of all switching surfaces along the x_1 direction is $x_1 \in \{-\gamma^2, 0, \gamma^2\}$. Thus, taking into consideration both summing terms, i.e. the full effect of $f(\mathbf{x})$ it follows that:

$$f_2 = \begin{cases} -\gamma^2 - \gamma & \text{if } \mathbf{x} \in G_{11}; \\ \gamma^2 + \gamma & \text{if } \mathbf{x} \in G_{12}; \\ \gamma^2 - \gamma & \text{if } \mathbf{x} \in G_{21}; \\ \gamma^2 + \gamma & \text{if } \mathbf{x} \in G_{22}. \end{cases} \tag{3.114}$$

By assuming $m = 3$ the elements in the partition G are double as when $m = 2$, now $G = \{G_{111}, G_{112}, G_{121}, G_{122}, G_{211}, G_{212}, G_{221}, G_{222}\}$ and:

$$f_2 = \begin{cases} -\gamma^3 - \gamma^2 - \gamma & \text{if } \mathbf{x} \in G_{111}; \\ -\gamma^3 - \gamma^2 + \gamma & \text{if } \mathbf{x} \in G_{112}; \\ -\gamma^3 + \gamma^2 - \gamma & \text{if } \mathbf{x} \in G_{121}; \\ -\gamma^3 + \gamma^2 + \gamma & \text{if } \mathbf{x} \in G_{122}; \\ \gamma^3 - \gamma^2 - \gamma & \text{if } \mathbf{x} \in G_{211}; \\ \gamma^3 - \gamma^2 + \gamma & \text{if } \mathbf{x} \in G_{212}; \\ \gamma^3 + \gamma^2 - \gamma & \text{if } \mathbf{x} \in G_{221}; \\ \gamma^3 + \gamma^2 + \gamma & \text{if } \mathbf{x} \in G_{222}. \end{cases} \tag{3.115}$$

The switching surfaces are located along the x_1 direction at

$$\begin{aligned} &-\gamma^3 - \gamma^2, \\ &-\gamma^3 + \gamma^2, \\ &\gamma^3 - \gamma^2, \\ &\gamma^3 + \gamma^2. \end{aligned} \tag{3.116}$$

Similarly, when $m = 4$ the elements in the partition G are double as when $m = 3$, now $G = \{G_{1111}, G_{1112}, G_{1121}, G_{1122}, G_{1211}, \ldots, G_{2221}, G_{2222}\}$ and:

$$f_2 = \begin{cases} -\gamma^4 - \gamma^3 - \gamma^2 - \gamma & \text{if } \mathbf{x} \in G_{1111}; \\ -\gamma^4 - \gamma^3 - \gamma^2 + \gamma & \text{if } \mathbf{x} \in G_{1112}; \\ -\gamma^4 - \gamma^3 + \gamma^2 - \gamma & \text{if } \mathbf{x} \in G_{1121}; \\ -\gamma^4 - \gamma^3 + \gamma^2 + \gamma & \text{if } \mathbf{x} \in G_{1122}; \\ -\gamma^4 + \gamma^3 - \gamma^2 - \gamma & \text{if } \mathbf{x} \in G_{1211}; \\ -\gamma^4 + \gamma^3 - \gamma^2 + \gamma & \text{if } \mathbf{x} \in G_{1212}; \\ -\gamma^4 + \gamma^3 + \gamma^2 - \gamma & \text{if } \mathbf{x} \in G_{1221}; \\ -\gamma^4 + \gamma^3 + \gamma^2 + \gamma & \text{if } \mathbf{x} \in G_{1222}; \\ \gamma^4 - \gamma^3 - \gamma^2 - \gamma & \text{if } \mathbf{x} \in G_{2111}; \\ \gamma^4 - \gamma^3 - \gamma^2 + \gamma & \text{if } \mathbf{x} \in G_{2112}; \\ \gamma^4 - \gamma^3 + \gamma^2 - \gamma & \text{if } \mathbf{x} \in G_{2121}; \\ \gamma^4 - \gamma^3 + \gamma^2 + \gamma & \text{if } \mathbf{x} \in G_{2122}; \\ \gamma^4 + \gamma^3 - \gamma^2 - \gamma & \text{if } \mathbf{x} \in G_{2211}; \\ \gamma^4 + \gamma^3 - \gamma^2 + \gamma & \text{if } \mathbf{x} \in G_{2212}; \\ \gamma^4 + \gamma^3 + \gamma^2 - \gamma & \text{if } \mathbf{x} \in G_{2221}; \\ \gamma^4 + \gamma^3 + \gamma^2 + \gamma & \text{if } \mathbf{x} \in G_{2222}. \end{cases} \tag{3.117}$$

The switching surfaces are located along the x_1 direction at:

$$
\begin{aligned}
&-\gamma^4 - \gamma^3 - \gamma^2, \\
&-\gamma^4 - \gamma^3 + \gamma^2, \\
&-\gamma^4 + \gamma^3 - \gamma^2, \\
&-\gamma^4 + \gamma^3 + \gamma^2, \\
&\gamma^4 - \gamma^3 - \gamma^2, \\
&\gamma^4 - \gamma^3 + \gamma^2, \\
&\gamma^4 + \gamma^3 - \gamma^2, \\
&\gamma^4 + \gamma^3 + \gamma^2.
\end{aligned}
\tag{3.118}
$$

Thus the effect of f_2 is that for each increment in m from m to $m+1$ the elements of the previous partition are doubled. The number of elements of the partition G is equal to 2^m.

Now, to analyze the another summing term of $f(\mathbf{x})$, f_1 can be rewritten in terms of f_2 as follows:

$$
f_1 = \alpha u \left(2 \left(x_1 - \sum_{i=1}^{m} w_i \right) - x_3,\ x_3 \right) = \alpha u \left(2(x_1 - f_2) - x_3,\ x_3 \right).
\tag{3.119}
$$

From this new form, it is easy to see that f_1 is basically a scaling of a step function $u(x, z)$. It is also possible to see that f_1 generates divides each element of the partition G in two elements with a surface of the form $\{\mathbf{x} \in \mathbb{R}^3, \epsilon \in \mathbb{R} : 2x_1 - x_3 = \epsilon\}$ to generate a new partition $P = \{P_1, \ldots, P_{2^{m+1}}\}$. This orientation of the switching surfaces generated by f_1 allows the existence of heteroclinic loops between the equilibria of adjacent elements of the partition P P_i and P_{i+1}, with $i = 1, 3, 5, \ldots, 2^{m+1} - 1$ which are related to the chaotic self-excited double-scroll attractors and the number of equilibria is 2^{m+1}. Therefore, for an appropriate selection of parameters 2^m self-excited attractors are generated.

Then, for instance, when $m = 1$ there are two self-excited attractors and one hidden attractor around these self-excited attractors, when m is incremented by one the number of attractors, self-excited and hidden included are doubled, but due to the symmetry, the two hidden attractors lead to the emergence of a new hidden attractor. Thus, generalizing this idea, the total number of hidden attractors is $2^m - 1$ and the total number of self-excited attractors is 2^m. This means that the number of coexisting attractors is $2^m + 2^m - 1 = 2^{m+1} - 1$.

The equilibria is located on the x_1 axis and its location depends on the parameters α, m and γ. Table 3.1 shows the location of the equilibria for

Table 3.1 Location of the equilibria at the x_1 axis for $m \in \{1, 2, 3\}$.

m	Location at x_1
$m = 1$	$-\gamma - \alpha, -\gamma + \alpha, \gamma - \alpha, \gamma + \alpha$
$m = 2$	$-\gamma^2 - \gamma - \alpha, -\gamma^2 - \gamma + \alpha, -\gamma^2 + \gamma - \alpha, -\gamma^2 + \gamma + \alpha,$ $\gamma^2 - \gamma - \alpha, \gamma^2 - \gamma + \alpha, \gamma^2 + \gamma - \alpha, \gamma^2 + \gamma + \alpha$
$m = 3$	$-\gamma^3 + \gamma^2 - \gamma - \alpha, -\gamma^3 + \gamma^2 - \gamma + \alpha, -\gamma^3 + \gamma^2 + \gamma - \alpha, -\gamma^3 + \gamma^2 + \gamma + \alpha,$ $-\gamma^3 - \gamma^2 - \gamma - \alpha, -\gamma^3 - \gamma^2 - \gamma + \alpha, -\gamma^3 - \gamma^2 + \gamma - \alpha, -\gamma^3 - \gamma^2 + \gamma + \alpha,$ $\gamma^3 + \gamma^2 - \gamma - \alpha, \gamma^3 + \gamma^2 - \gamma + \alpha, \gamma^3 + \gamma^2 + \gamma - \alpha, \gamma^3 + \gamma^2 + \gamma + \alpha,$ $\gamma^3 - \gamma^2 - \gamma - \alpha, \gamma^3 - \gamma^2 - \gamma + \alpha, \gamma^3 - \gamma^2 + \gamma - \alpha, \gamma^3 - \gamma^2 + \gamma + \alpha$

three cases, $m \in \{1, 2, 3\}$. The self-excited attractors oscillate around a pair of equilibria $\mathbf{x}^*_{eq_k}$ and $\mathbf{x}^*_{eq_{k+1}}$, with $k = 1, 3, \ldots, 2^{m+1} - 1$.

It is important to note that γ must be large enough to allow the emergence of the hidden attractors. As a first example of this approach, consider the case with $a = 0.2$, $b = 5$, $c = -3$, $\alpha = 1$, $\gamma = 10$ and $m = 1$, the system exhibits two self-excited attractors and one hidden attractor as shown in the Fig. 3.37.

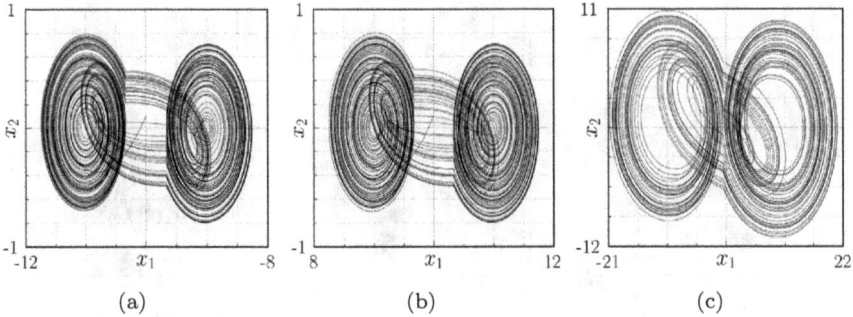

Fig. 3.37 System given by (3.100), (3.101), (3.102) and (3.103) with $a = 0.2$, $b = 5$, $c = -3$, $\alpha = 1$, $\gamma = 10$ and $m = 1$. (a) Self-excited double-scroll at $[-10, 0, 0]^T$ with $X_0 = [-10, 0.1, 0]^T$ and $t \in [0, 200]$. (b) Self-excited double-scroll at $[10, 0, 0]^T$ with $X_0 = [10, 0.1, 0]^T$ and $t \in [0, 200]$. (c) Hidden double-scroll attractor with $X_0 = [0, 0, 0]^T$ and $t \in [10000, 10100]$.

Now let us consider an increment in m, thus $m = 2$ with the same parameters $a = 0.2$, $b = 5$, $c = -3$, $\alpha = 1$, $\gamma = 10$. Now the system exhibits two hidden attractors separated in a symmetrical way from the $x_2 - x_3$ plane, each of the hidden attractors oscillates around two self excited attractors and there is also a new hidden attractor oscillating around the two hidden attractors as shown in the Fig. 3.38

Table 3.2   Lyapunov exponents of the fifteen attractors of the system given by (3.100), (3.101), (3.102) and (3.103) with $a = 0.2$, $b = 5$, $c = -3$, $\alpha = 1$, $\gamma = 10$ and $m = 2$.

Attractor	Center position at x_1	$x(0)$	Lyapunov exponents
1	-1110.0	$(-1110, 0.05, 0)^T$	0.343,-0.0,-2.745
2	-1090.0	$(-1090, 0.05, 0)^T$	0.343,0.0,-2.744
3	-910.0	$(-910, 0.05, 0)^T$	0.348,-0.0,-2.744
4	-890.0	$(-890, 0.05, 0)^T$	0.347,0.0,-2.744
5	890.0	$(890, 0.05, 0)^T$	0.35,-0.0,-2.745
6	910.0	$(910, 0.05, 0)^T$	0.346,0.0,-2.743
7	1090.0	$(1090, 0.05, 0)^T$	0.346,0.0,-2.741
8	1110.0	$(1110, 0.05, 0)^T$	0.341,0.0,-2.739
9	-1100.0	$(-1100, 0.5, 0)^T$	0.306,-0.0,-2.509
10	-900.0	$(-900, 0.5, 0)^T$	0.31,0.0,-2.51
11	900.0	$(900, 0.5, 0)^T$	0.31,-0.0,-2.51
12	1100.0	$(1100, 0.5, 0)^T$	0.31,0.0,-2.51
13	-1000.0	$(-1000, 5.0, 0)^T$	0.35,-0.02,-2.46
14	1000.0	$(1000, 5.0, 0)^T$	0.35,-0.02,-2.46
15	0.0	$(0, 50.0, 0)^T$	0.36,-0.03,-2.43

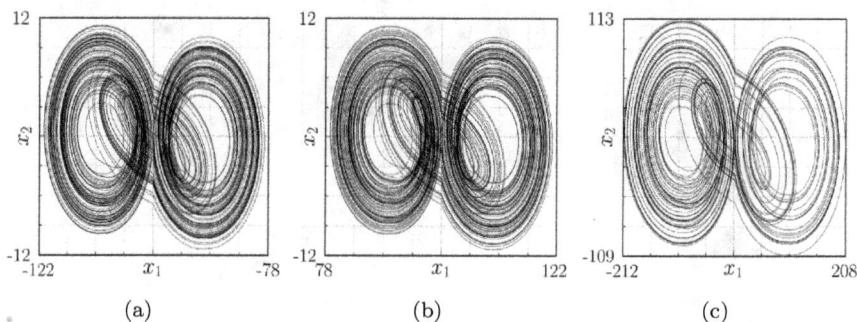

(a)   (b)   (c)

Fig. 3.38   System given by (3.100), (3.101), (3.102) and (3.103) with $a = 0.2$, $b = 5$, $c = -3$, $\alpha = 1$, $\gamma = 10$ and $m = 2$. (a) Hidden double-scroll attractor at $[-100, 0, 0]^T$ with $X_0 = [-100, 0.1, 0]^T$ and $t \in [0, 200]$. (b) Hidden double-scroll attractor at $[100, 0, 0]^T$ with $X_0 = [100, 0.1, 0]^T$ and $t \in [0, 200]$. (c) Hidden double-scroll attractor with $X_0 = [0, 0, 0]^T$ and $t \in [10000, 10100]$.

Finally let us consider $a = 0.2$, $b = 5$, $c = -3$, $\alpha = 1$, $\gamma = 10$ and $m = 3$. The system exhibits fifteen attractors, eight self-excited and seven hidden attractors, the tree larger hidden attractors are shown in the Fig. 3.39. All the attractors are chaotic according to the computed Lyapunov spectra shown in Table 3.2.

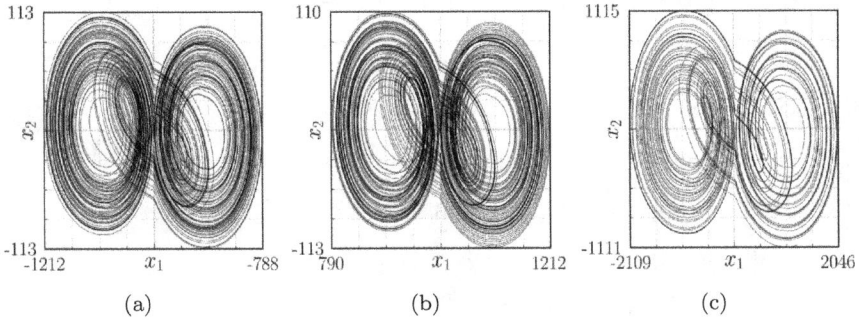

Fig. 3.39 System given by (3.100), (3.101), (3.102) and (3.103) with $a = 0.2$, $b = 5$, $c = -3$, $\alpha = 1$, $\gamma = 10$ and $m = 3$. (a) Hidden double-scroll attractor at $[-1000, 0, 0]^T$ with $X_0 = [-1000, 0.1, 0]^T$ and $t \in [0, 200]$. (b) Hidden double-scroll attractor at $[1000, 0, 0]^T$ with $X_0 = [1000, 0.1, 0]^T$ and $t \in [0, 200]$. (c) Hidden double-scroll attractor with $X_0 = [0, 0, 0]^T$ and $t \in [10000, 10100]$.

3.7 Generation of pseudo random numbers based on hidden attractors

In this section, some of the previously presented systems with hidden attractors are tested with the approach reported in [García-Martínez *et al.* (2015)] to explore if it is feasible to use them in the generation of pseudo random numbers. The approach consist in the numerical integration of a solution of the system which lead to n points of the form $\mathbf{x_i} = [x_{1i}, x_{2i}, x_{3i}]^T$, then n numbers k_i are found as follows:

$$k_i = \left\lfloor \sum_{i=1}^{3} x_{1i} \cdot 10^{14} \right\rfloor \mathrm{mod} 256. \qquad (3.120)$$

Thus, $k_i \in [0, 255]$ for $i = 1, \ldots, n$. Since each k_i can be represented by eight bits, each k_i generates a string of eight zeros and ones. This $8n$ zeros and ones are saved into a file which would be used in the test suite provided by the National Institute of Standards and Technology (NIST) in order to verify if the approach along with the system used are good to be used as a pseudo random number generator.

The test suite performs fifteen statistical test:

(1) The Frequency (Monobit) Test,

(2) Frequency Test within a Block,

(3) The Runs Test,

(4) Tests for the Longest-Run-of-Ones in a Block,

(5) The Binary Matrix Rank Test,

(6) The Discrete Fourier Transform (Spectral) Test,

(7) The Non-overlapping Template Matching Test,

(8) The Overlapping Template Matching Test,

(9) Maurer's "Universal Statistical" Test,

(10) The Linear Complexity Test,

(11) The Serial Test,

(12) The Approximate Entropy Test,

(13) The Cumulative Sums (Cusums) Test,

(14) The Random Excursions Test, and,

(15) The Random Excursions Variant Test.

However, some tests are decomposable into a variety of subtests.

The interpretation of empirical results can be conducted in any number of ways. Two approaches NIST has adopted include (1) the examination of the proportion of sequences that pass a statistical test and (2) the distribution of P-values to check for uniformity. The test suites generates a file named finalAnalysisReport.txt that contains the results of these two forms of analysis [Rukhin (2010)].

Each performed test throws a P-value reflected in the generated file. The P-value, frequently called the "tail probability" is the probability (under the null hypothesis of randomness) that the chosen test statistic will assume values that are equal to or worse than the observed test statistic value when considering the null hypothesis [Rukhin (2010)].

In the experiments performed in this section the number of binary sequences under test are $m = 500$ and the significance level $\alpha = 0.1$. An $\alpha = 0.01$ indicates that one would expect one sequence in 100 sequences to be rejected by the test if the sequence was random. NIST recommends to fix the significance level to be at least 0.001, but no larger than 0.01 [Rukhin (2010)].

Thus, if for instance $\alpha = 0.01$ and 499 binary sequences had P-values ≥ 0.01, then the proportion is $499/500 = 0.998$. The range of acceptable proportions is determined using the confidence interval defined by the

[Rukhin (2010)] as

$$\hat{p} = \pm\sqrt{\frac{\hat{p}(1-\hat{p})}{m}} \tag{3.121}$$

where $\hat{p} = 1 - \alpha$. Thus, in the experiments of this section, $m = 500$, $\alpha = 0.01$, which gives the values:

$$0.99 + 0.0133491572767722 = 1.003349157276772$$
$$0.99 - 0.0133491572767722 = 0.9766508427232278. \tag{3.122}$$

The class of systems used in this section along with the PRNG algorithm is the introduced in Section 3.2 given by (3.40) and (3.44) with $a = 1$, $b = 10$, $v = 5$, $\alpha = 1$, $\gamma = 10$, $d = 0.1$ and $m \in \{1, 2, 3, 4, 9, 49, 99\}$.

The system exhibits hidden scroll attractor with $m + 1$ scrolls centered at $[0, 0, 0]^T$. Thus, to generate the 1000 the initial conditions has been selected according to te following:

$$\begin{aligned} X_0 &= [x_{10}, x_{20}, x_{30}]^T, \\ x_{10} &\in [-0.5, 0.5], \\ x_{20} &= 0, \\ x_{30} &= 0. \end{aligned} \tag{3.123}$$

For the case $m = 1$ th scroll attractor is shown in the Fig. 3.40(a). The results are shown in the Fig. 3.40(b) and the p-values are shown in the Table 3.3.

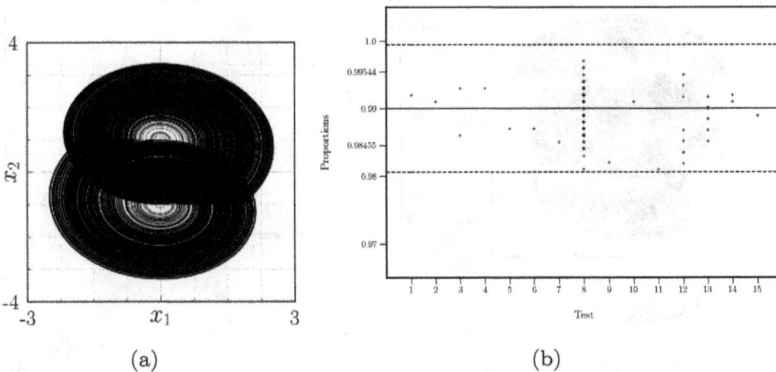

(a)   (b)

Fig. 3.40   (a) Hidden scroll attractor exhibited by the system given by (3.40) and (3.44) with $a = 1$, $b = 10$, $v = 5$, $\alpha = 1$, $\gamma = 10$, $d = 0.1$ and $m = 1$. (b) Test results.

For the case $m = 2$ the scroll attractor is shown in Fig. 3.41(a). The results are shown in the Fig. 3.41(b) and the p-values are shown in the Table 3.3.

Table 3.3   Test results from the fifteen test when using the hidden scroll attractor exhibited by the system given by (3.40) and (3.44) with $a = 1$, $b = 10$, $v = 5$, $\alpha = 1$, $\gamma = 10$, $d = 0.1$ and $m = 1$.

Test No.	Test name	Portion with P-value $\geq \alpha$	Result F=Fail, S=Success
1	Frequency	0.992	S
2	BlockFrequency	0.991	S
3	CumulativeSums	0.986	S
4	CumulativeSums	0.993	S
5	Runs	0.993	S
6	LongestRun	0.987	S
7	Rank	0.987	S
8	FFT	0.985	S
9	NonOverlappingTemplate	0.9895	S
10	OverlappingTemplate	0.982	S
11	Universal	0.991	S
12	ApproximateEntropy	0.981	S
13	RandomExcursions	0.9876	S
12	RandomExcursionsVariant	0.9900	S
13	Serial	0.991	S
14	Serial	0.992	S
15	LinearComplexity	0.989	S

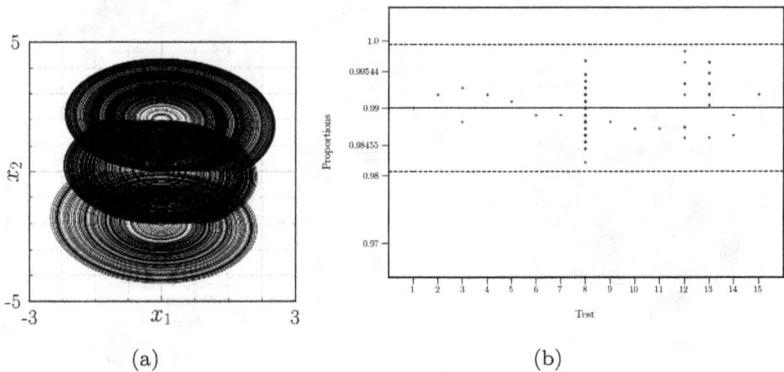

Fig. 3.41   (a) Hidden scroll attractor exhibited by the system given by (3.40) and (3.44) with $a = 1$, $b = 10$, $v = 5$, $\alpha = 1$, $\gamma = 10$, $d = 0.1$ and $m = 2$. (b) Test results.

For the case $m = 2$ the scroll attractor is shown in the Fig. 3.42(a). The results are shown in the Fig. 3.42(b) and the p-values are shown in the Table 3.5.

Table 3.4   Test results from the fifteen test when using the hidden scroll attractor
exhibited by the system given by (3.40) and (3.44) with $a = 1$, $b = 10$, $v = 5$,
$\alpha = 1$, $\gamma = 10$, $d = 0.1$ and $m = 2$.

Test No.	Test name	Portion with P-value $\geq \alpha$	Result F=Fail, S=Success
1	Frequency	0.99	S
2	BlockFrequency	0.992	S
3	CumulativeSums	0.988	S
4	CumulativeSums	0.993	S
5	Runs	0.992	S
6	LongestRun	0.991	S
7	Rank	0.989	S
8	FFT	0.989	S
9	NonOverlappingTemplate	0.9897	S
10	OverlappingTemplate	0.988	S
11	Universal	0.987	S
12	ApproximateEntropy	0.987	S
13	RandomExcursions	0.9918	S
12	RandomExcursionsVariant	0.9904	S
13	Serial	0.989	S
14	Serial	0.986	S
15	LinearComplexity	0.992	S

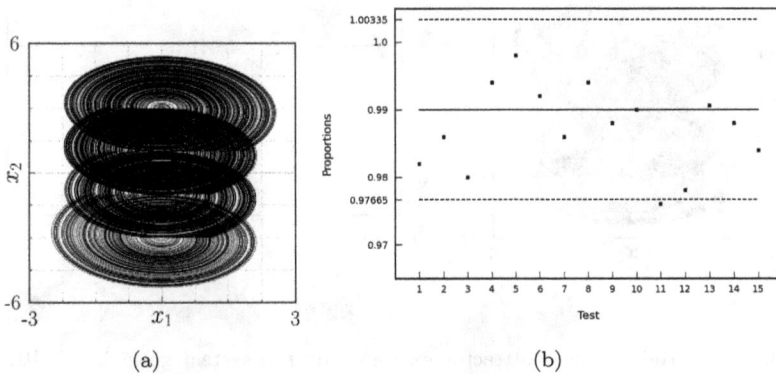

Fig. 3.42   (a) Hidden scroll attractor exhibited by the system given by (3.40) and (3.44) with $a = 1$, $b = 10$, $v = 5$, $\alpha = 1$, $\gamma = 10$, $d = 0.1$ and $m = 3$. (b) Test results.

For the case $m = 2$ the scroll attractor is shown in the Fig. 3.43(a).
The results are shown in the Fig. 3.43(b) and the p-values are shown in the
Table 3.6.

Table 3.5   Test results from the fifteen test when using the hidden scroll attractor exhibited by the system given by (3.40) and (3.44) with $a = 1$, $b = 10$, $v = 5$, $\alpha = 1$, $\gamma = 10$, $d = 0.1$ and $m = 2$.

Test No.	Test name	Portion with P-value $\geq \alpha$	Result F=Fail, S=Success
1	Frequency	0.992	S
2	BlockFrequency	0.984	S
3	CumulativeSums	0.99	S
4	Runs	0.992	S
5	LongestRun	0.996	S
6	Rank	0.988	S
7	FFT	0.994	S
8	NonOverlappingTemplate	0.998	S
9	OverlappingTemplate	0.986	S
10	Universal	0.99	S
11	ApproximateEntropy	0.988	S
12	RandomExcursions*	0.990	S
13	RandomExcursionsVariant*	0.990	S
14	Serial	0.986	S
15	LinearComplexity	0.99	S

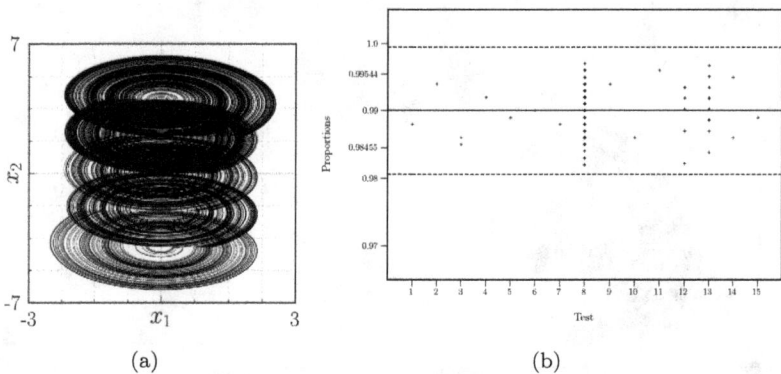

(a)   (b)

Fig. 3.43   (a) Hidden scroll attractor exhibited by the system given by (3.40) and (3.44) with $a = 1$, $b = 10$, $v = 5$, $\alpha = 1$, $\gamma = 10$, $d = 0.1$ and $m = 4$. (b) Test results.

For the case $m = 2$ the scroll attractor is shown in the Fig. 3.44(a). The results are shown in the Fig. 3.44(b) and the p-values are shown in the Table 3.7.

For the case $m = 2$ the scroll attractor is shown in the Fig. 3.45(a). The results are shown in the Fig. 3.45(b) and the p-values are shown in the Table 3.8.

Table 3.6   Test results from the fifteen test when using the hidden scroll attractor exhibited by the system given by (3.40) and (3.44) with $a = 1$, $b = 10$, $v = 5$, $\alpha = 1$, $\gamma = 10$, $d = 0.1$ and $m = 4$.

Test No.	Test name	Portion with P-value $\geq \alpha$	Result F=Fail, S=Success
1	Frequency	0.992	S
2	BlockFrequency	0.984	S
3	CumulativeSums	0.99	S
4	Runs	0.992	S
5	LongestRun	0.996	S
6	Rank	0.988	S
7	FFT	0.994	S
8	NonOverlappingTemplate	0.998	S
9	OverlappingTemplate	0.986	S
10	Universal	0.99	S
11	ApproximateEntropy	0.988	S
12	RandomExcursions*	0.990	S
13	RandomExcursionsVariant*	0.990	S
14	Serial	0.986	S
15	LinearComplexity	0.99	S

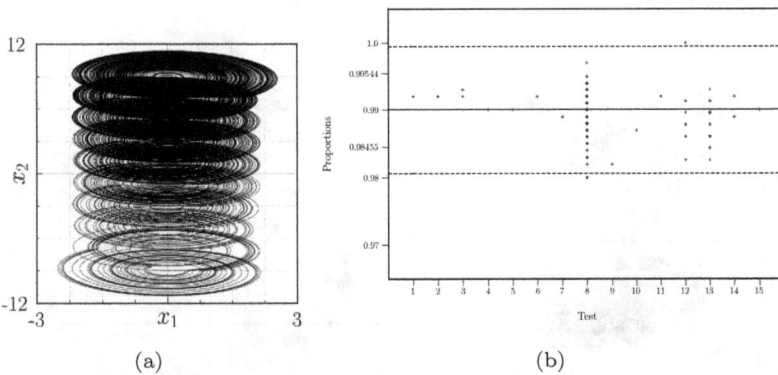

(a)   (b)

Fig. 3.44   (a) Hidden scroll attractor exhibited by the system given by (3.40) and (3.44) with $a = 1$, $b = 10$, $v = 5$, $\alpha = 1$, $\gamma = 10$, $d = 0.1$ and $m = 9$. (b) Test results.

For the case $m = 2$ the scroll attractor is shown in the Fig. 3.46(a). The results are shown in the Fig. 3.46(b) and the p-values are shown in the Table 3.9.

The results show that when the number of scrolls is increased in the hidden attractor the results from the NIST test are worse and some test

Table 3.7 Test results from the fifteen test when using the hidden scroll attractor exhibited by the system given by (3.40) and (3.44) with $a = 1$, $b = 10$, $v = 5$, $\alpha = 1$, $\gamma = 10$, $d = 0.1$ and $m = 9$.

Test No.	Test name	Portion with P-value $\geq \alpha$	Result F=Fail, S=Success
1	Frequency	0.992	S
2	BlockFrequency	0.992	S
3	CumulativeSums	0.993	S
4	CumulativeSums	0.992	S
5	Runs	0.99	S
6	LongestRun	0.99	S
7	Rank	0.992	S
8	FFT	0.989	S
9	NonOverlappingTemplate	0.9896	S
10	OverlappingTemplate	0.982	S
11	Universal	0.987	S
12	ApproximateEntropy	0.992	S
13	RandomExcursions	0.9889	S
12	RandomExcursionsVariant	0.9861	S
13	Serial	0.989	S
14	Serial	0.992	S
15	LinearComplexity	0.99	S

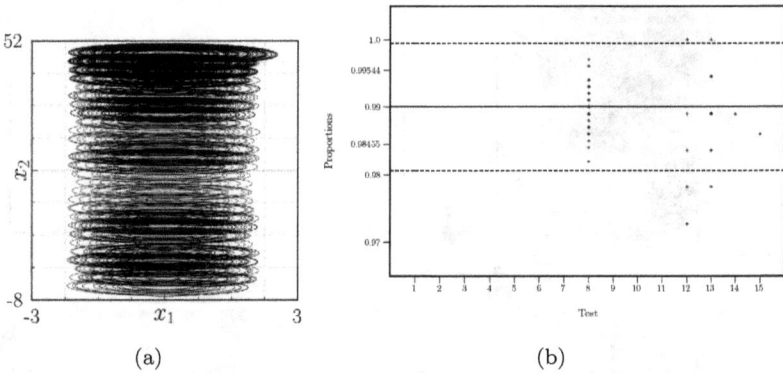

(a)   (b)

Fig. 3.45 (a) Hidden scroll attractor exhibited by the system given by (3.40) and (3.44) with $a = 1$, $b = 10$, $v = 5$, $\alpha = 1$, $\gamma = 10$, $d = 0.1$ and $m = 49$. (b) Test results.

are failed. Thus, the hidden double-scroll attractor is recommended for cryptographic applications over a scroll attractor with higher number of scrolls.

Table 3.8 Test results from the fifteen test when using the hidden scroll attractor exhibited by the system given by (3.40) and (3.44) with $a = 1$, $b = 10$, $v = 5$, $\alpha = 1$, $\gamma = 10$, $d = 0.1$ and $m = 49$.

Test No.	Test name	Portion with P-value $\geq \alpha$	Result F=Fail, S=Success
1	Frequency	0.155	F
2	BlockFrequency	0.156	F
3	CumulativeSums	0.153	F
4	CumulativeSums	0.151	F
5	Runs	0.17	F
6	LongestRun	0.237	F
7	Rank	0.183	F
8	FFT	0.186	F
9	NonOverlappingTemplate	0.4419	F
10	OverlappingTemplate	0.204	F
11	Universal	0.222	F
12	ApproximateEntropy	0.19	F
13	RandomExcursions	0.9836	S
12	RandomExcursionsVariant	0.9890	S
13	Serial	0.244	F
14	Serial	0.989	S
15	LinearComplexity	0.986	S

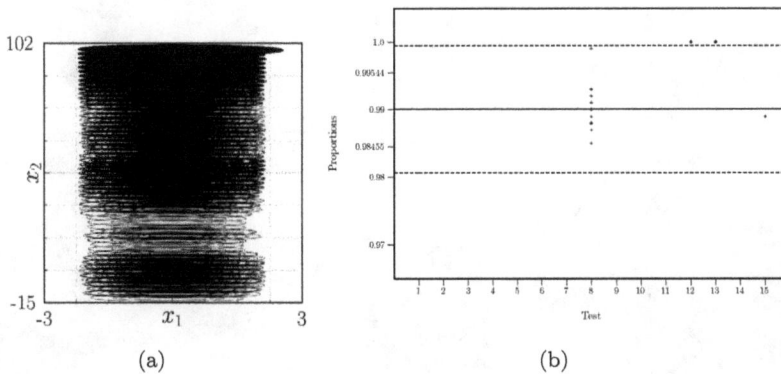

(a)   (b)

Fig. 3.46 (a) Hidden scroll attractor exhibited by the system given by (3.40) and (3.44) with $a = 1$, $b = 10$, $v = 5$, $\alpha = 1$, $\gamma = 10$, $d = 0.1$ and $m = 99$. (b) Test results.

Table 3.9   Test results from the fifteen test when using the hidden scroll attractor exhibited by the system given by (3.40) and (3.44) with $a = 1$, $b = 10$, $v = 5$, $\alpha = 1$, $\gamma = 10$, $d = 0.1$ and $m = 99$.

Test No.	Test name	Portion with P-value $\geq \alpha$	Result F=Fail, S=Success
1	Frequency	0.013	F
2	BlockFrequency	0.015	F
3	CumulativeSums	0.014	F
4	CumulativeSums	0.013	F
5	Runs	0.013	F
6	LongestRun	0.019	F
7	Rank	0.015	F
8	FFT	0.017	F
9	NonOverlappingTemplate	0.2033	F
10	OverlappingTemplate	0.018	F
11	Universal	0.018	F
12	ApproximateEntropy	0.017	F
13	RandomExcursions	0.9916	S
12	RandomExcursionsVariant	1.0	S
13	Serial	0.02	F
14	Serial	0.188	F
15	LinearComplexity	0.989	S

Chapter 4

Fractional-order PWL systems

4.1 Fractional dynamical systems theory

Non-integer derivatives and integrals are appropriate for modeling systems complex. Fractional-order models have been shown to be more suitable to model systems in which the properties of non-locality are present, heritability, self-similarity and pseudo-randomness such as: mechanical problems inverse, stochastic kinetics, deterministic chaos, movement in viscous fluids, between others, an example of this type of phenomenon is Brownian motion.

4.1.1 *Fractional operators*

Derivatives and integrals of fractional-order are generalizations of those of integer order. However, in the literature we can find a variety of different definitions for fractional derivatives [Diethelm (2010); Podlubny (1998); Monje *et al.* (2010); Petráš (2011)], the most used being the Riemann-Liouville and Caputo definitions [Diethelm (2010)]. The Riemann-Liouville fractional derivative is defined as:

$$D_a^n f(x) = \frac{1}{\Gamma(m-n)} \frac{d^m}{dx^m} \int_a^x \frac{f(t)}{(x-t)^{n-m+1}} dt, \qquad (4.1)$$

and the Caputo definition is described by:

$$D_0^n f(x) = \frac{1}{\Gamma(m-n)} \int_a^x \frac{f^{(m)}(t)}{(x-t)^{n-m+1}} dt, \qquad (4.2)$$

with $n \in \mathbb{R}$, m the ceil function of n $(m = \lceil n \rceil)$, and Γ is the gamma function which is defined as:

$$\Gamma(z) = \int_0^\infty t^{z-1}e^{-t}dt. \tag{4.3}$$

The Riemann-Liouville fractional derivative played a determining role in the development of the body of theory for fractional calculus, and was used successfully in strictly mathematical applications. But when trying to carry out the mathematical modeling of real physical phenomena by means of fractional differential equations, the problem of the initial conditions also of fractional-order arose. These types of conditions are not physically interpretable and present a considerable obstacle when it comes to making practical use of fractional calculus. The Caputo differential operator, in contrast to the Riemann-Liouville operator, uses as initial conditions derivatives of integer order, which is why the study operator used in this work is the Caputo operator.

4.1.2   *Existence and uniqueness for Caputo's fractional differential equations*

Corresponding to the Caputo operator, in this section we will discuss the problem of existence and uniqueness of solutions.

Consider the equation of the form:

$$D_0^{n_k}x(t) = f(t, x(t)), \tag{4.4}$$

subject to initial conditions

$$x^{(j)}(0) = x_0^{(j)}, \text{ with } j = 0, 1, \ldots, \lceil n_k \rceil - 1.$$

With respect to the existence of the solution, the first correct result It responds to Peano's existence theorem for first-order equations.

Theorem 4.1. *Let* $0 < n$ *and* $m = \lceil n \rceil$. *Also let* $x_0^0, \ldots, x_0^{(m-1)} \in \mathbb{R}$, $K > 0$ *and* $h^* > 0$. *Define* $G := \{(t, x) : t \in [0, h^*].|x - \sum_{k=0}^{m-1} t^k x_0^{(k)}/k!| \leq K\}$, *and be te funtion* $f : G \to \mathbb{R}$ *continuous. Additionally define* $M := sup_{(t,z) \in G}|f(t,z)|$ *and*

$$h := \begin{cases} h^*, & if\ M = 0; \\ min\{h^*.(K\Gamma(n+1)/M)^{1/n}\}, & other\ case. \end{cases}$$

Then, exist one function $x \in C[0, h]$ *that fulfil the initial value problem* (4.4).

Lemma 4.1. *Assume the hypotheses of the Theorem 4.1. The function* $x \in C[0,h]$ *is a solution of the initial value problem* (4.4) *if and only if it is a solution of the nonlinear Volterra integral equation of the second kind*

$$x(t) = \sum_{k=0}^{m-1} \frac{t^k}{k!} x_0^{(k)} + \frac{1}{\Gamma(n)} \int_0^t (x - \tau)^{n-1} f(\tau, x(\tau)) d\tau. \qquad (4.5)$$

It is important to note that under certain assumptions, the solution exists on the entire interval $[0, h^*]$ (and for all t for which $f(t,x)$ is defined) and not just for the subinterval $[0, h]$ with some $h \leq h^*$.

We now turn to the theorem on uniqueness of solutions which corresponds to the well-known Picard-Lindel result.

Theorem 4.2. *Let* $0 < n$ *and* $m = \lceil n \rceil$. *Also let* $x_0^0, \ldots, x_0^{(m-1)} \in \mathbb{R}$, $K > 0$ *and* $h^* > 0$. *Define the set* G *according to the Theorem 4.1 and let the function* $f : G \to \mathbb{R}$ *be continuous and Lipschitz with respect to a second variable, that is:*

$$|f(t, x_1) - f(t, x_2)| \leq L|x_1 - x_2|, \qquad (4.6)$$

with a constant $L > 0$ *independent of* t, x_1 *and* x_2. *Then, denoting* h *as in the Theorem 4.1, there exists a unique function* $y \in C[0, h]$ *that solves the initial value problem* (4.4).

In the case of linear systems it is necessary to consider another theorem for existence and uniqueness. In which a slightly different set of assumptions are considered that allow us to arrive at a result of existence and uniqueness for a broader class of systems including linear equations.

Theorem 4.3. *Let* $0 < n$ *and* $m = \lceil n \rceil$. *Also let* $x_0^0, \ldots, x_0^{(m-1)} \in \mathbb{R}$ *and* $h^* > 0$. *Define the set* $G := [0, h^*] \times \mathbb{R}$ *and let the function* $f : G \to \mathbb{R}$ *be continuous and satisfy the Lipschitz condition with respect to a second variable with a Lipschitz constant* $L > 0$ *which is independent of* t, x_1 *and* x_2. *Then there is a unique defined function* $x \in C[0, h^*]$ *that solves the initial value problem* (4.4).

The immediate consequence is obtained:

Corollary 4.1. *Let* $0 < n$ *and* $m = \lceil n \rceil$. *Also let* $x_0^0, \ldots, x_0^{(m-1)} \in \mathbb{R}$ *and* $h^* > 0$. *Define the set* $G := [0, \infty) \times \mathbb{R}$ *and let the function* $f : G \to \mathbb{R}$

be continuous and satisfy the Lipschitz condition with respect to a second variable with a Lipschitz constant $L > 0$ which is independent of t, x_1 and x_2. Then there is a unique defined function $x \in C[0, \infty]$ that solves the initial value problem (4.4).

Now we can find the explicit solution of a system for a simple class of fractional differential equations of Caputo type, called homogeneous linear equations with constant coefficients.

Theorem 4.4. *Let $0 < n$ and $m = \lceil n \rceil$ and $\lambda \in \mathbb{R}$. The solution of the initial value problem*

$$D_0^n x(t) = \lambda x(t), \qquad x(0) = x_0, \quad x^k(0) = 0 \quad (k = 1, 2, \ldots, m-1), \quad (4.7)$$

is given by:

$$x(t) = x_0 E_n(\lambda t^n), \qquad t \geq 0, \tag{4.8}$$

with E_n the Mittag-Leffler function.

Definition 4.1. Let $n > 0$. The function E_n defined by:

$$E_n(z) := \sum_{j=0}^{\infty} \frac{z^j}{\Gamma(jn+1)}. \tag{4.9}$$

Whenever the series converges, is called the Mittag-Leffler function of order n.

The differential equation considered in the Theorem 4.4 is a simple example of a linear fractional differential equation. For proofs of the results presented in this section see reference [Diethelm (2010)].

4.1.3   *Fractional-order dynamical systems*

Theorem 4.5. *[Diethelm (2010)] A general time-invariant system of commensurate fractional-order is described as follows:*

$$D_0^{n_k} x(t) = f(t, x(t), D_0^{n_1} x(t), D_0^{n_2} x(t), \ldots, D_0^{n_{k-1}} x(t)), \tag{4.10}$$

subject to initial conditions

$$x^{(j)}(0) = x_0^{(j)}, \quad with \ j = 0, 1, \ldots, \lceil n_k \rceil - 1,$$

where $n_1, n_2, \ldots, n_k \in \mathbb{Q}$, such that $n_k > n_{k-1} > \cdots > n_1 > 0$, $n_j - n_{j-1} \leq 1$ for for all $j = 2, 3, \ldots, k$ and $0 < n_1 \leq 1$, let M be the least common multiple of the denominator of $n_1, n_2, \ldots, n_k$ and let $\alpha = 1/M$ and $N = Mn_k$.

Then the Eq. (4.10) is equivalent to the following system of equations

$$D_0^\alpha x_0(t) = x_1(t),$$
$$D_0^\alpha x_1(t) = x_2(t),$$
$$\vdots \qquad\qquad\qquad\qquad (4.11)$$
$$D_0^\alpha x_{N-2}(t) = x_{N-1}(t),$$
$$D_0^\alpha x_{N-1}(t) = f(t, x_0(t), x_{n1/\alpha}(t), \ldots, x_{n_{k-1}/\alpha}(t)),$$

with initial conditions

$$x_j(0) = \begin{cases} x_0^{(j/M)} & \text{if } j/M \in \mathbb{N} \cup \{0\}, \\ 0 & \text{other case.} \end{cases}$$

Furthermore, according to the Theorem 4.5, this time-invariant linear system can be expressed in matrix form as follows:

$$\frac{d^\alpha \mathbf{x}(t)}{dt^\alpha} = A\mathbf{x}, \qquad (4.12)$$

where $\mathbf{x} \in \mathbb{R}^n$ is the vector of states, $A \in \mathbb{R}^{n \times n}$ is a linear operator, and α is the commensurate order of derivative of fractional-order $0 < \alpha < 1$.

In fractional-order systems, the region of stability depends on the order of the derivative α as represented in Fig. 4.1. It is important to note that the stability of the equilibrium point can be controlled by means of the derivative order α, for example, a hyperbolic saddle point of an integer order system can be transformed to a stable equilibrium point by change the order of derivative α of the system.

This is an important consideration for the design of chaotic attractors since the interest is in unstable dynamics.

The stability of fractional-order systems is stated as follows:

- **Asymptotically stable:** The system (4.12) is asymptotically stable if and only if $|arg(\lambda)| > \frac{\alpha\pi}{2}$ for all eigenvalues λ of matrix A. In this case, the solution $x(t) \to 0$ as $t \to \infty$.

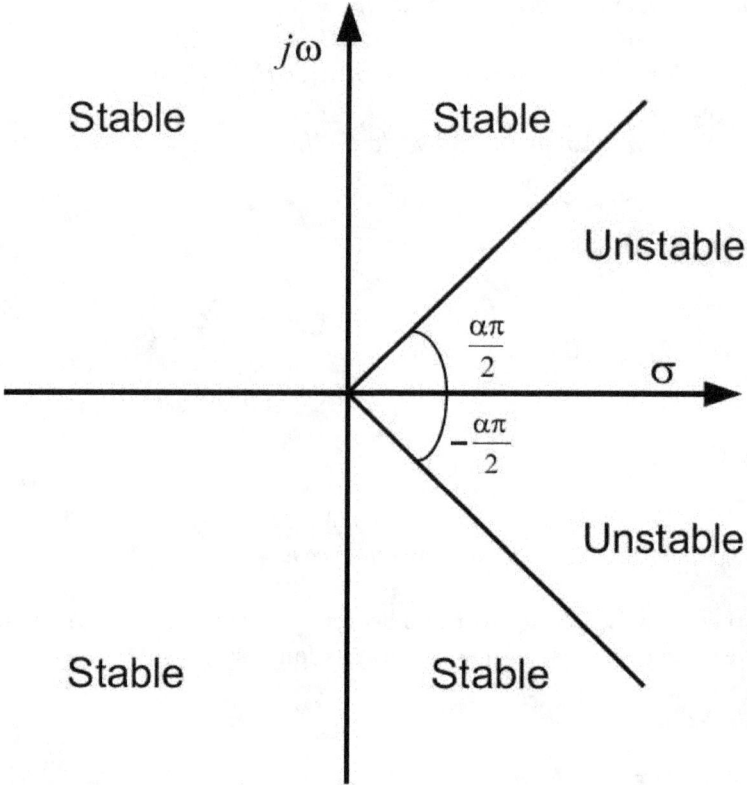

Fig. 4.1   Region of stability for a system of differential equations of fractional-order $0 < \alpha < 1$.

- **Stable:** The system (4.12) is stable if and only if $|arg(\lambda)| \geq \frac{\alpha\pi}{2}$ for all eigenvalues λ of matrix A obeying that the critical eigenvalues must satisfy $|arg(\lambda)| = \frac{\alpha\pi}{2}$ and have a geometric multiplicity of one.

The interest is to have unstable dynamics in order to obtain chaotic attractors, then the system is restricted to have at least one eigenvalue in the unstable region, that is, the system (4.12) is unstable if and only if $|arg(\lambda)| < \frac{\alpha\pi}{2}$ for at least one of its eigenvalues λ of matrix A.

Additionally, the system given by (4.12) with its equilibrium point at the origin can be generalized by an affine linear system as follows:

$$\frac{d^\alpha \mathbf{x}(t)}{dt^\alpha} = A\mathbf{x} + B, \tag{4.13}$$

where $B \in \mathbb{R}^n$ is a constant vector and $A \in \mathbb{R}^{n \times n}$ is a non-singular linear operator. Now the equilibrium point $p \equiv (x_1^*, x_2^*, \ldots, x_N^*)^T = -A^{-1}B$ of a commensurate general affine linear system of fractional-order (4.13) with fractional-order $0 < \alpha < 1$, is a saddle equilibrium point if its eigenvalues $\lambda_1, \lambda_2, \ldots, \lambda_\kappa, \lambda_{\kappa+1}, \ldots, \lambda_n$ of the Jacobian matrix evaluated at the equilibrium point satisfies the following condition.

$$|arg(\lambda_i)| > \frac{\alpha\pi}{2} \quad \text{with } i = 1, 2, \ldots, \kappa,$$
$$|arg(\lambda_i)| < \frac{\alpha\pi}{2} \quad \text{with } i = \kappa + 1, \kappa + 2, \ldots, n. \tag{4.14}$$

Note that we are interested in working with unstable systems, that is, systems that do not meet the asymptotic local stability condition.

$$min|arg(\lambda_i)| > \frac{\alpha\pi}{2}, \text{ for } i = 1, 2, \ldots, n. \tag{4.15}$$

4.1.4 *Numerical method for solving fractional differential equations*

There are no methods that can provide, analytically, the exact solution of any fractional differential equation like those provided by integer order systems, therefore it is necessary to use numerical methods. The Adams-Bashforth-Moulton (ABM) method, a corrective predictor scheme, was reported in [Diethelm (2010)] and is used to obtain the temporal evolution of fractional systems. The algorithm is a generalization of the classical Adams-Bashforth-Moulton integrator which is known for solving first-order problems such as switched systems [Zambrano-Serrano *et al.* (2016)].

Now we present the method which is widely used and has been proven efficient in many practical applications. [Connolly and Ford (2006); Tavazoei *et al.* (2009)].

Consider the fractional differential equation that is described by (4.10) as follows:

$$D^\alpha x(t) = f(t, x(t)), \quad 0 \le t \le T;$$
$$x^{(k)}(0) = x_0^{(k)}, \quad k = 0, 1, \ldots, n - 1. \tag{4.16}$$

We assume the function f is such that a unique solution exists in an interval $[0, T]$, and we assume that we work on a uniform mesh $\{t_j = jh : j = 0, 1, \ldots, N\}$ with some integer N and $h = T/N$.

The solution of (4.16) is given by an integral equation of the Volterra type as:

$$x(t) = \sum_{k=0}^{\lceil \alpha \rceil - 1} x_0^k \frac{t^k}{k!} + \frac{1}{\Gamma(\alpha)} \int_0^t (t-z)^{\alpha-1} f(z, x(z)) dz, \tag{4.17}$$

$$x_{k+1} = \sum_{k=0}^{\lceil \alpha \rceil - 1} x_0^k \frac{t^k}{k!} + \frac{1}{\Gamma(\alpha)} \left(\sum_{j=0}^k a_{j,k+1} f(t_j, x_j) + a_{k+1,k+1} f(t_{k+1}, x_{k+1}^P) \right), \tag{4.18}$$

where

$$a_{j,k+1} = \begin{cases} \dfrac{h^\alpha}{\alpha(\alpha+1)} (k^{\alpha+1} - (k-\alpha)(k+1)^\alpha), & \text{if } j = 0; \\[2ex] \dfrac{h^\alpha}{\alpha(\alpha+1)} ((k-j+2)^{\alpha+1} + (k-j)^{\alpha+1} \\ -2(k-j+1)^{\alpha+1}), & \text{if } 1 \le j \le k; \\[2ex] \dfrac{h^\alpha}{\alpha(\alpha+1)}, & \text{if } j = k+1; \end{cases} \tag{4.19}$$

with the predictor scheme given as follows:

$$x_{k+1}^P = x(0) + \frac{1}{\Gamma(n)} \sum_{j=0}^k b_{j,k+1} f(t_j, x_j), \tag{4.20}$$

and

$$b_{j,k+1} = \frac{h^\alpha}{\alpha} ((k+1-j)^\alpha - (k-j)^\alpha). \tag{4.21}$$

The error of this approximation is given by

$$max_{j=0,1,\dots,N} |x(t_j) - x_h(t_j)| = O(h^p). \tag{4.22}$$

It is important to mention that in this chapter, models of fractional-order will be considered. Therefore, the algorithm presented above is used to obtain the numerical solutions of the fractional differential equations presented in the book.

4.2 Approximation of integer order systems to fractional-order systems

It was previously mentioned that fractional-order models have been shown to be more suitable to model systems with certain characteristics, or phenomena with certain uncertainties, as a example consider the equation for a single pendulum:

$$\frac{d^2\theta}{dt^2} = \frac{-g}{L}sen(\theta) \tag{4.23}$$

with its linear approximation as:

$$\frac{d^2\theta}{dt^2} = \frac{-g}{L}\theta \tag{4.24}$$

or making $\theta = \theta_1$, and $\theta_2 = \dot{\theta}_1$:

$$\begin{pmatrix} \dot{\theta}_1 \\ \dot{\theta}_2 \end{pmatrix} = \begin{pmatrix} 0 & \theta_2 \\ \frac{-g}{L}\theta_1 & 0 \end{pmatrix}. \tag{4.25}$$

The solution of Eq. (4.25) is shown in Figs. 4.2(a)–(b), the evolution of θ with respect to time and the system solution into the phase portrait, respectively. For this case, that is an idealization of the model we can see that the oscillations never end, which does not happen in reality. We know that by adding a term related with friction into the equation would help us to have a better representation of reality, instead we will consider that the derivative is of non-integer order as follows:

$$\begin{pmatrix} D^q\theta_1 \\ D^q\theta_2 \end{pmatrix} = \begin{pmatrix} 0 & \theta_2 \\ \frac{-g}{L}\theta_1 & 0 \end{pmatrix} \tag{4.26}$$

where D^q is the Caputo fractional-order operator. The solution of fractional system (4.26) is presented in Figs. 4.2(c)–(d), the evolution of θ with respect to time and the system solution into the phase portrait, respectively. For this case the system evolution looks more like what is observed in reality.

4.3   Scroll attractors

For scroll attractors, a fractional-order unstable dissipative system (FOUDS) will be defined, considering the dynamical characteristics of the integer-order UDS that have been previously defined, the interest is in found a minimum efective order while chaos behavior is preserved in the first case. The fractional chaotic attractor appears as a result of the combination of several unstable one-spiral trajectories around a saddle hyperbolic stationary point. In the proposed fractional-order chaotic system the resulting fractional chaotic attractor has same equilibria than scroll as shown herein. The fractional chaos state could be verified using the time series analysis of the numerical temporal data.

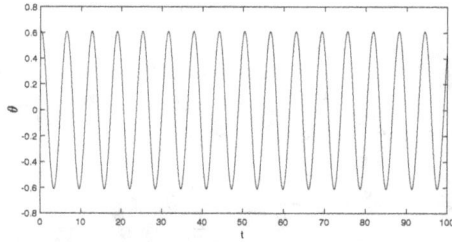

(a)

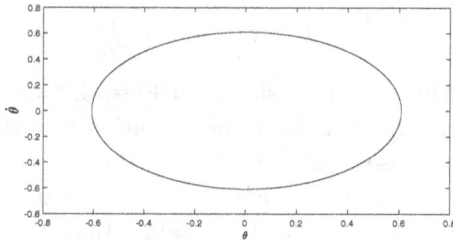

(b)

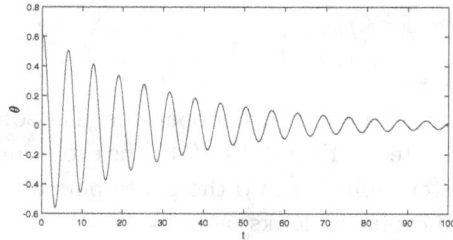

(c)

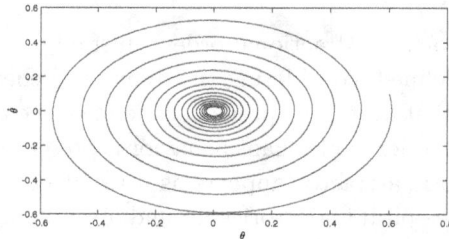

(d)

Fig. 4.2   Trajectory projections for the system given by Eq. (4.26), (a), (b) with $q = 1.0$ angular position and the solution on the phase space, respectively. (c), (d) with $q = 0.98$ angular position versus time and the solution on the phase space.

Consider a multi-scroll generator system based on the jerk equation [Gilardi-Velázquez *et al.* (2017); Echenausía-Monroy *et al.* (2018); Campos-Cantón (2016); Echenausía-Monroy and Huerta-Cuellar (2019)], which fulfill all the conditions to be considered as a UDS I for the parameters a_i [Gilardi-Velázquez *et al.* (2017)] as follows:

$$D^q x = y,$$
$$D^q y = z, \tag{4.27}$$
$$D^q z = -a_1 x - a_2 y - a_3 z - f(x),$$

where D^q is the Caputo fractional-order operator. The parameters a_i are a constant values that defines the eigenvalues of the system, where $a_i \in \mathbb{R}^+$. It is possible to generate an attractor with multiple scrolls by the use of a commutation law $f(x)$, for our case is used the Nearest Integer Function $Round(x)$, as shown in [Gilardi-Velázquez *et al.* (2017); Huerta-Cuellar *et al.* (2014a)], whose purpose is to divide the state space into subdomains D_i of the same size, and add the system equilibrium points, being achieved through the coexistence of a large number of one-spiral unstable trajectories.

Since the systems described in Eq. (4.30) can be represented in the form of Eq. (4.13) the local stability for each equilibrium point is defined by the eigenvalues of linear operator A, whose characteristic polynomial is $\lambda^3 + a_3\lambda^2 + a_2\lambda + a_1$. According with Proposition 2.2, for $0 < a_1$, $0 < a_2 < a_1/a_2$, $0 < a_3$ the system is UDS type I and correspond to all of the equilibrium points.

On the other hand, the local stability region of the fractional-order system depends of two things principally, the system eigenvalues spectra, related with the operating region of the UDS i.e. the a_i parameters of Eq. (4.30), and the second one integration order used since modify the stability region, fact because of the multi-scroll behavior will be analyzed under a fractional-order of $0.93 \le q \le 1$, in which it is contained the critical integration order (q_c) with which it is possible to analyze the multi-scroll generator, for fixed parameters defined later, without becoming stable, when all the state variables are considered with an equal integration order.

By considering a system based on the jerk equation (4.30), as a $f(x)$ the Round function as:

$$f(x) = C1 * Round(x/C2), \tag{4.28}$$

and for the following parameter values: $a_1 = 10.5$; $a_2 = 7$; $a_3 = 0.7$; $C1 = 0.9$ and $C2 = 0.6$. It is worth to mention that these parameter values are

the same to those used in [Gilardi-Velázquez *et al.* (2017)]. With which
it is possible to generate an attractor with multiple scrolls by means of a
commutation law, in this case Eq. (4.28), this induces an infinity of commutation surfaces, and thus guarantee the generation of the same number
of scrolls as space divisions. In Figure 4.3 the trajectory of the system
Eq. (4.30) for $q = 0.96$ and the aforementioned parameters is displayed, in
the phase space (a), onto the $x - y$, $x - z$ and $y - z$ planes (b), (c) and (d),
respectively.

The local behavior near the equilibrium point is determined by the
Jacobian of the system (4.30) which has the following spectrum $\Lambda = \{\lambda_1 = -0.8304, \lambda_2 = 0.3651 + 1.2935i, \lambda_3 = 0.3651 - 1.2935i\}$. This spectrum
determines the critical value of derivative order $\alpha_c \approx 0.8249$ to get the
system (4.30) to be stable,

$$q < \frac{2}{\pi} min|arg(\lambda_i)| \approx 0.8249. \tag{4.29}$$

Consider D_i as each domain defined by the *Round* function and a
Poincaré planes implemented exactly at the commutation surface around
the origin generated by this function, Poincaré sections S_1, S_2 are implemented at $x = 0.3$ and $x = -0.3$ respectively, according with the commutation values defined. A trajectory in D_i, with equilibrium point $X_i^* \in \mathcal{D}_i$.
When the trajectory oscillates around X_i^* increases the distance to the equi-
librium point and exits the current domain $\mathcal{D}_i$ to $\mathcal{D}_{i+1}$ in Fig. 4.4(a) (or $\mathcal{D}_i$
to $\mathcal{D}_{i-1}$ in Fig. 4.4(b)) through the commutation surface near the region of
intersection of the unstable manifold with the Poincaré plane, as it can be
appreciated with the intersection points of the trajectory ϕ^{t_j} in Fig. 4.4,
marked in green cross the points that represent the crossing points from left
to right and in blue cross the from right to left for each plane considered.
Notice that most of the trajectories are distributed uniformly in both ways.

Now consider

$$\begin{aligned} D^{q_x}x &= y, \\ D^{q_y}y &= z, \\ D^{q_z}z &= -\alpha[x + y + z - f(x)], \end{aligned} \tag{4.30}$$

where D^{q_i} is the Caputo fractional-order operator, and $i = \{x, y, z\}$. The
control parameter α is a constant that defines the eigenvalues of the system,
where $\alpha \in \mathbb{R}$.

According with Proposition 2.1, the range of control parameters is
defined for the system analysis, the variation in the $\alpha \in [0.35, 0.75]$ for
$q = 0.96$.

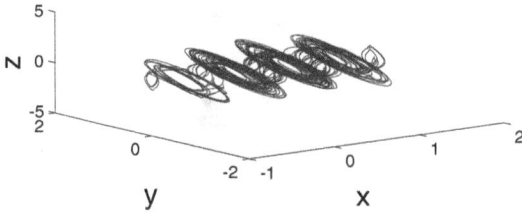

(a)

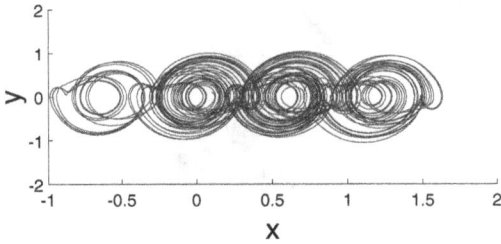

(b)

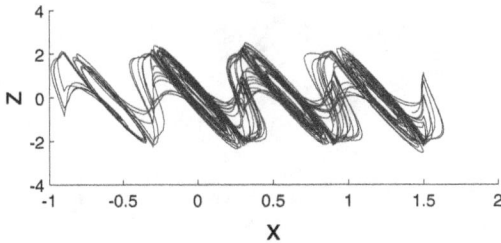

(c)

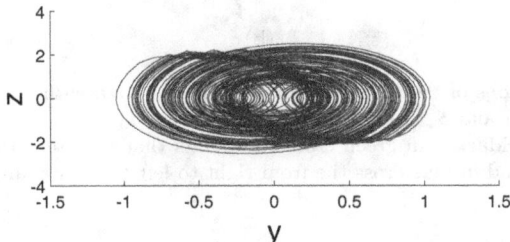

(d)

Fig. 4.3   Trajectory of the system (4.30) given by Eq. (4.28) (a) onto the phase space, in (b) (c) (d) projections onto the (x, y); (x, z); (y, z) respectively for $a_1 = 10.5$; $a_2 = 7$; $a_3 = 0.7$; $C1 = 0.9$; $C2 = 0.6$, $q = 1$ and $\alpha = 1$.

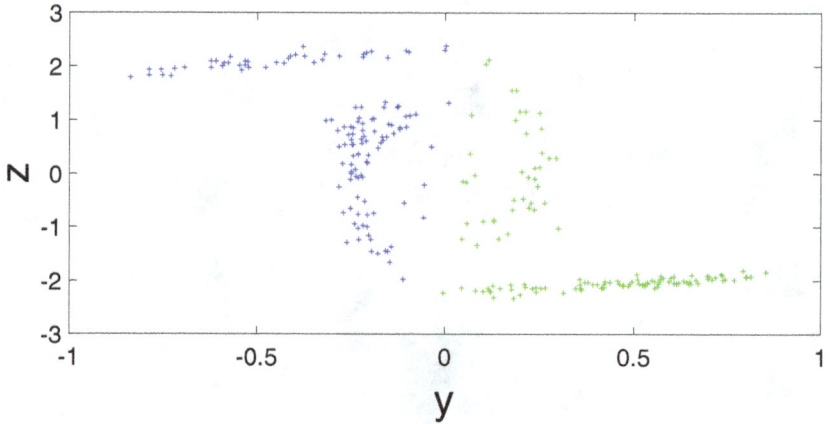

(a)

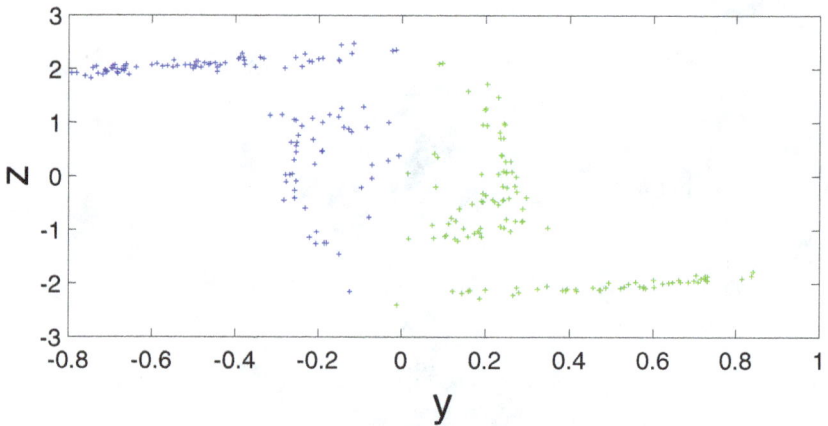

(b)

Fig. 4.4   Intersections of the trajectory of the system (4.30) with (4.28) with the commutation surface S_1 and S_2 (b) for $a_1 = 10.5$; $a_2 = 7$; $a_3 = 0.7$; $C1 = 0.9$; $C2 = 0.6$, $\alpha = 1$ and $q = 1$. Marked in green cross the points that represent the crossing points from left to right and in blue cross the from right to left for each plane.

Five analysis scenarios are determined, shown in Fig. 4.5. Considering S_l as each scenario, where $q^T = [q_x,\ q_y,\ q_z] = S_l$, for $l = 1, 2, \ldots, 5$, as the integration order vector applied in Eq. (4.30), being q_i, for $i = \{x, y, z\}$, the integration order of each state variable in the system, the analysis scenarios

(a)

(b)

(c)

(d)

(e)

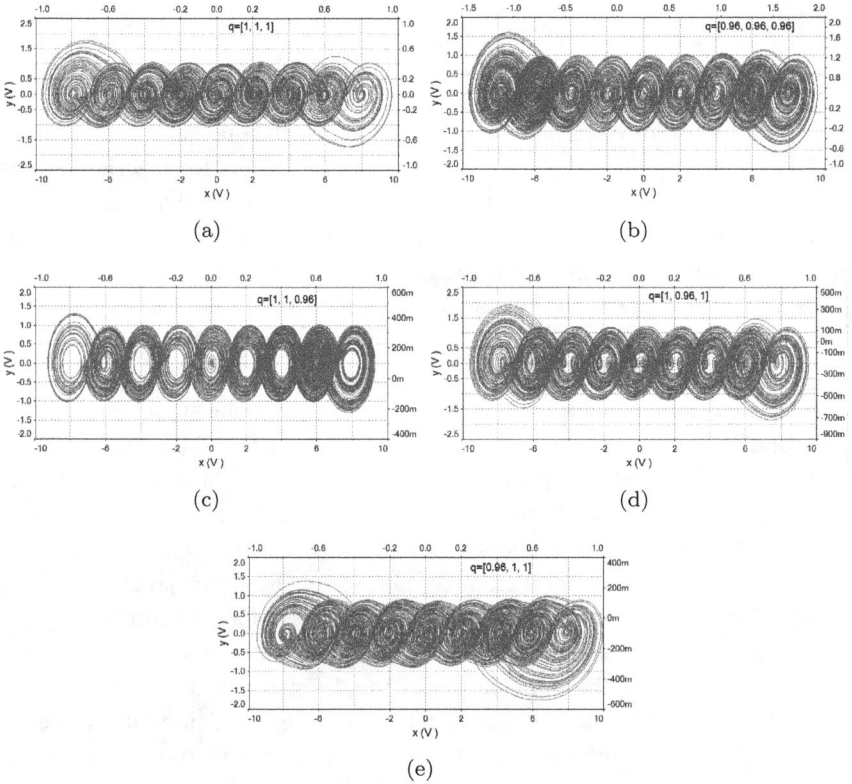

Fig. 4.5 Attractors generated by the system described in Eq. (4.30) for the control parameter $\alpha = 0.45$ and integration order: (a) $q^T = [1, 1, 1]$, (b) $q^T = [0.96, 0.96, 0.96]$, (c) $q^T = [1, 1, 0.96]$, (d) $q^T = [1, 0.96, 1]$, (e) $q^T = [0.96, 1, 1]$.

are defined as:

- $S1 = [1, \, 1, \, 1]$.
- $S2 = [0.96, \, 0.96, \, 0.96]$.
- $S3 = [1, \, 1, \, 0.96]$.
- $S4 = [1, \, 0.96, \, 1]$.
- $S5 = [0.96, \, 1, \, 1]$.

4.4 Fractional multi-stable systems

According with reference [Echenausía-Monroy *et al.* (2020)], it is possible generate multi-stable behavior in PWL systems through fractional derivatives. Following the mechanism reported in [Echenausía-Monroy *et al.* (2020)], the system described by Eq. (4.30) is now analyzed numerically through the use of the Adams-Bashforth-Moulton (ABM) method [Diethelm *et al.* (2002)] for under changes in the integration order q. The algorithm were constructed as a generalization of the classical ABM integrator that is well known in the resolution of first-order switching systems problems [Zambrano-Serrano *et al.* (2016); Gilardi-Velázquez and Campos-Cantón (2018b)]. Considering the stability limitations for fractional-order systems, the system described for Eq. (4.30) have a critic integration-order at $q_c = 0.927$ to preserve the local instability. Following the stability theory for fractional systems [Zambrano-Serrano *et al.* (2016)], if the system is analyzed with integration orders $q < q_c$, the dynamics of the system will be stable, turning each equilibrium point into a focus attractor.

The fractional-order system described in Eq. (4.30), is analyzed for a fixed bifurcation parameter $\alpha = 1.0$, and by consider the switching function described by $Round(x)$, an integration step size of $h = 0.01$, and considering the same integration order in the three-state variables. As in previous examples, a bifurcation diagram of fractional-order versus local maximum in the x state variable is depicted in Fig. 4.6(a), where it can be seen the qualitative changes in the system dynamics as the derivative order is reduced, this behavior is similar to that calculated for the integer system. Furthermore, in Fig. 4.6(b) shows the number of domains $\mathcal{D}_i$ that the system visit considering the same values of the derivative order q. Notice that, if the range of $q \leq 0.945$ approximately, the number of domains $\mathcal{D}$ visited remain at the constant value of 3, because the system presents only one scroll located in the domain around origin as depicted in Fig. 4.7.

As in the case for the integer order system under changes in α, in Figs. 4.7(b) and (d) present a projection of one trajectory onto the plane (x, y) and (y, z) both initialized with the same initial condition near the origin $\mathbf{X}_0 = (0.1, 0.1, 0)^T$. Notice in the projections that as the time increases the oscillating clockwise system trajectory grows larger, until eventually the trajectories on the scrolls cross the commutation surface. Apparently both systems trajectory crosses the commutation surface plane S_1 and S_2 in a neighborhood near to the intersection of the unstable and stable manifolds. By observing different projections of the attractor, for example the projection onto the planes (x, z), in Figs. 4.7(c) and (d), it can be appreciated that the trajectory has returns in the direction of equilibrium point associated to the domain, around $x = \pm 0.3$ in Figs. 4.7(b), (c) and $y = \pm 0.5$, $z = \pm 3.0$. Which has been slightly changed due to the variation of the parameter from $q = 1$ to $q = 0.945$.

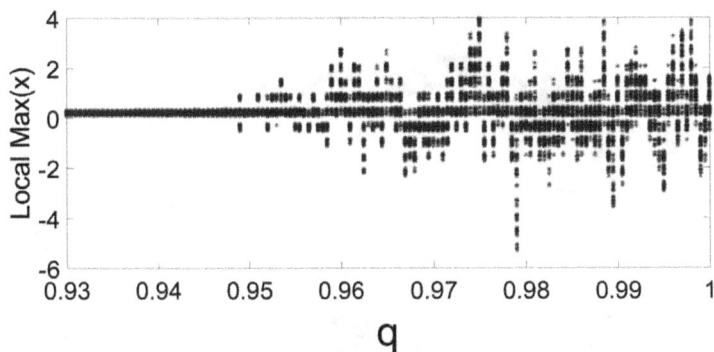

(a)

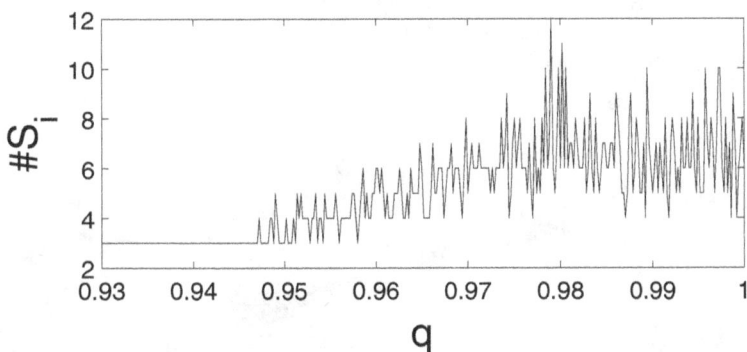

(b)

Fig. 4.6 (a) Bifurcation diagram of the system given by Eqs. (4.30) with (4.28) for $a_1 = 10.5$; $a_2 = 7$; $a_3 = 0.7$; $C1 = 0.9$; $C2 = 0.6$, $\alpha = 1.0$, for the value of $0.93 \leq q \leq 1.0$. Figure (b) shows the number of domains $\mathcal{D}_i$ visited by the trajectory of the system for the same values of the bifurcation parameters above. The initial condition considered for both diagrams is $\mathbf{X}_0 = (0.1, 0.1, 0)$.

4.5 Effects and interpretation of using fractional operators

There are not many works developed in the aim of provide a interpretation of the effects in a dynamical system caused by the use of integration orders minors that the unit. There are few comparative analyzes that show, quantitatively, the changes that occurs in the dynamics of a system when contemplating fractional-orders, against the "natural" dynamics of the model when analyzed with an entire-integration order. In this section, the changes that the system undergoes when fractional-order is contem-

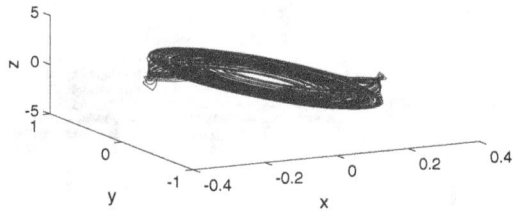

(a)

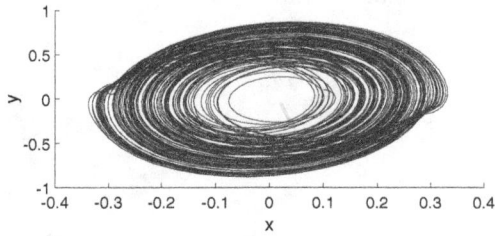

(b)

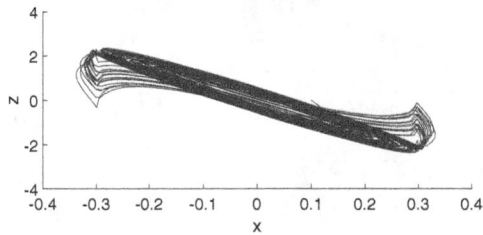

(c)

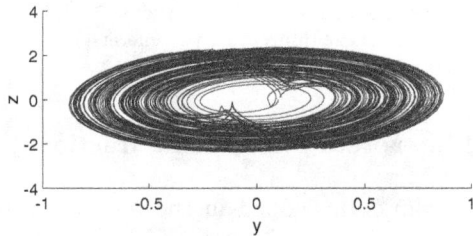

(d)

Fig. 4.7   Trajectory of the system (4.30) given by Eq. (4.28) (a) onto the phase space, in (b) (c) (d) projections onto the (x, y); (x, z); (y, z) respectively for $a_1 = 10.5$; $a_2 = 7$; $a_3 = 0.7$; $C1 = 0.9$; $C2 = 0.6$, $q = 0.945$ and $\alpha = 1$.

plated in the whole system, as independently in each of the state variables, are analyzed. Interpretation of the fractional derivatives effects in a jerk multi-scroll system is is mainly addressed.

4.5.1 *Frequency*

Some fractional analyzes have been reported in Rössler systems as in [Zhang *et al.* (2009)], for which numerical studies and characterizations have been performed. In this section, a characterization of the fractional-order equilibrium points of the Rössler system is shown numerically. Here, the conventional derivative is replaced by a fractional derivative, as follows:

$$\begin{aligned} Dx^q &= -wy - z, \\ Dy^q &= -wx + ay, \\ Dz^q &= b + z(x - c), \end{aligned} \qquad (4.31)$$

where w represents the natural frequency of the system [Pisarchik and Jaimes-Reátegui (2015)] and the location of the equilibrium points is modified as follows: $x1^* = (c + \sqrt{(c^2 - 4ab)})/(2), w(-c - \sqrt{(c^2 - 4ab)})/(2a), w^2(c + \sqrt{(c^2 - 4ab)})/(2a))$, and $x2^* = (c - \sqrt{(c^2 - 4ab)}, w(-c + \sqrt{(c^2 - 4ab)})/(2a), w^2(c - \sqrt{(c^2 - 4ab)})/(2a))$, note that the x coordinate does not depend on w. For $a = 0.16$ $b = 0.1$ $c = 8.5$, the projections of $x1^*$ and $x2^*$ onto the plane (y, z) are shown in Figs. 4.8(a) and (b), respectively.

To determine the stability of the equilibrium points, the eigenvalues associated with the Jacobian matrix of the system were calculated as a function of w. The real component associated with $x1^*$ is shown in Fig. 4.9(a) and the imaginary component is in Fig. 4.9(b).

Note that the real part of the eigenvalues is not changed by the changes in w and that for this case the three eigenvalues have a positive real part and two of them are complex conjugates. The eigenvalues belonging to $x2^*$ are shown in Fig. 4.10(a) the real part, and in Fig. 4.10(b) the imaginary component.

Through the use of fractional operator, it is possible to modify the local stability of the system, since the stability region in the complex plane depends on the derivative order q, as shown in Fig. 4.1. As mentioned earlier, the stability of the equilibrium point can be controlled using the derivative order q. For example, an unstable equilibrium point for an integer order system can be transformed into a stable point by the exact derivative order q of the system.

(a)

(b)

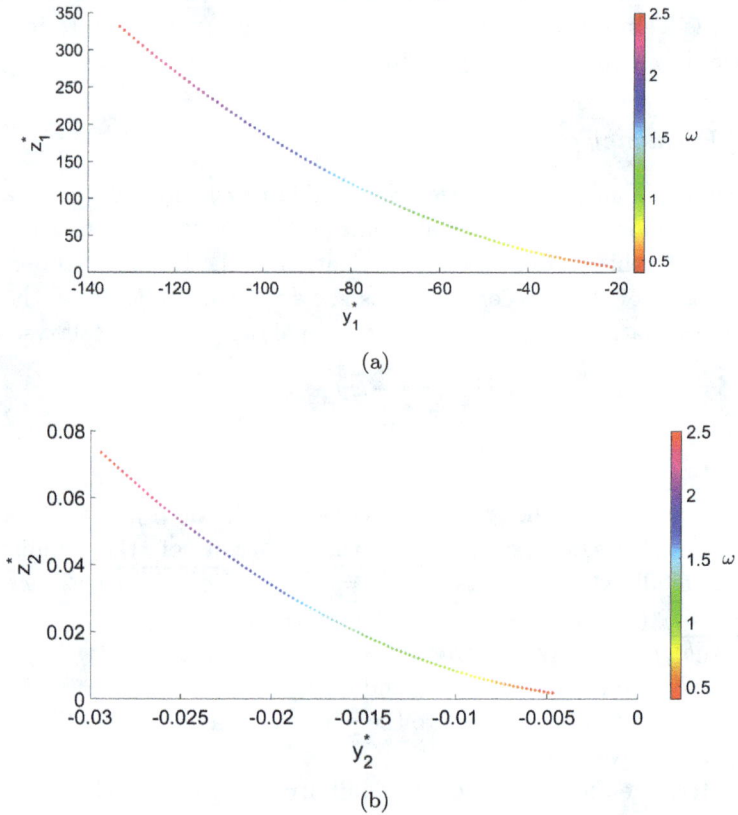

Fig. 4.8   Projection of the equilibria on the plane (y, z) as a function of w for: (a) $x1^*$, (b) $x2^*$.

The equilibrium point is asymptotically stable if and only if $|arg(\lambda)| > \frac{q\pi}{2}$ for all eigenvalues λ associated with A. In this case, the solution $x(t) \rightarrow 0$ as $t \rightarrow \infty$.

Considering that $x1^*$ has a positive real eigenvalue, it remains as an unstable equilibrium point despite the derivative order used. On the other hand, the stability of $x2^*$ could be modified by the derivative order, and since our interest is to obtain a non-stable behavior around this equilibrium point, Fig. 4.11 shows the minimum value of the derivative before the equilibrium point becomes stable.

In addition, the response frequency Ω of the system was analyzed as a function of w and the derivative order q. Figure 4.12 shows a map of the

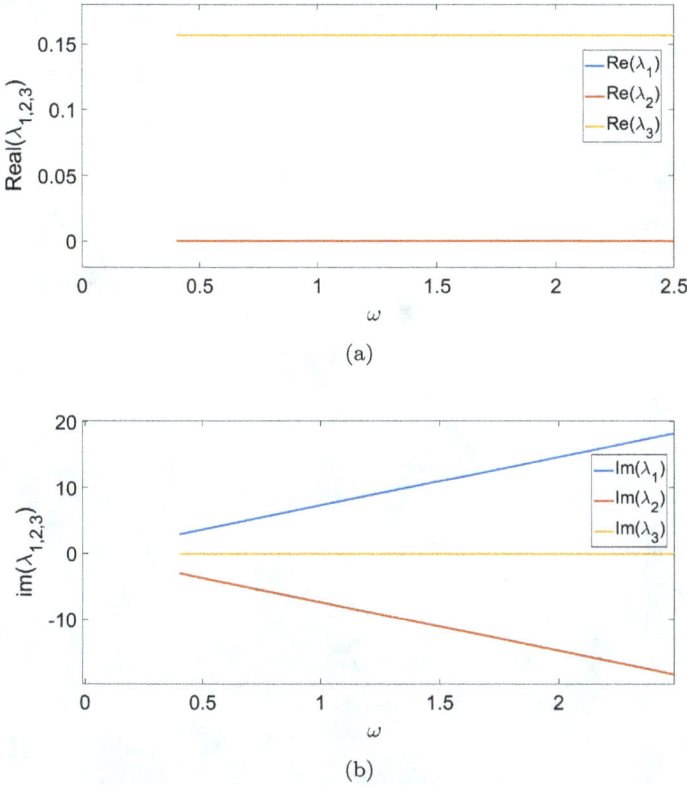

Fig. 4.9 (a) Real part of the eigenvalues associated to $x1^*$ versus w, imaginary part of the eigenvalues associated to $x1^*$ versus w (b).

response frequency Ω as a function of w and q, indicating that changing the derivative order has no effect on the dynamic frequency. For derivative values where the equilibrium point $x2^*$ is not stable, the system responds to the input frequency and this value remains practically constant.

4.5.2 *Poincaré sections analysis*

Poincaré planes are used to graphically represent the crossing points between the domains where the scroll are generated through the switching surfaces [Gilardi-Velázquez *et al.* (2017)]. A Poincaré planes were implemented exactly at the commutation surface as in the previous section. The crossing events of interest are to identify the crossing point, or in other words, the escape points and return points to the origin domain $\mathcal{D}_i$ to $\mathcal{D}_{i+1}$

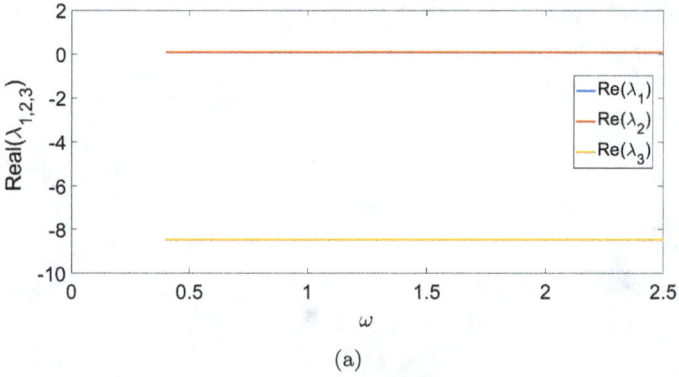

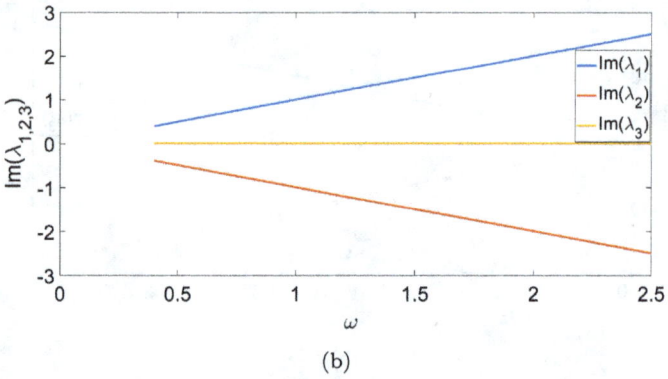

Fig. 4.10 (a) Real part of the eigenvalues associated to $x2^*$ versus w, imaginary part of the eigenvalues associated to $x2^*$ versus w (b).

or $\mathcal{D}_i$ to $\mathcal{D}_{i-1}$. In Fig. 4.14 it is shown the trajectory intersections for the system (4.30) with $q = 0.945$ for the right and left plane, respectively, in red asterisk the crosses from right to left and the black asterisk the crossing from left to right, in green and blue crosses the crossing points for $q = 1$ and $\alpha = 1$ described in the Section 4.4.

Since the goal of analyzing multi-stable behavior with a Poincaré section is to identify the relation between the integer and fractional system and analyze the behavior of each attractor at the switching surfaces, the crossing points in each plane for the multi-stable attractors obtained. The results are shown in Figs. 2.28, 4.6, 4.13 and 4.14. The bifurcation analysis show a great similarity in the evolution of the dynamics under the parametric changes considered in each case. Nevertheless, in the analysis

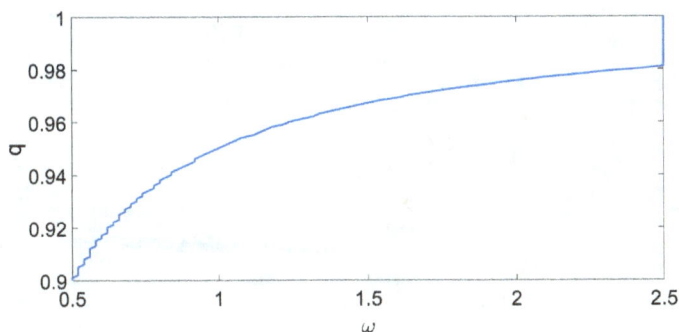

Fig. 4.11   Minimum value of derivative order q, before the equilibrium point becomes stable, in function of w.

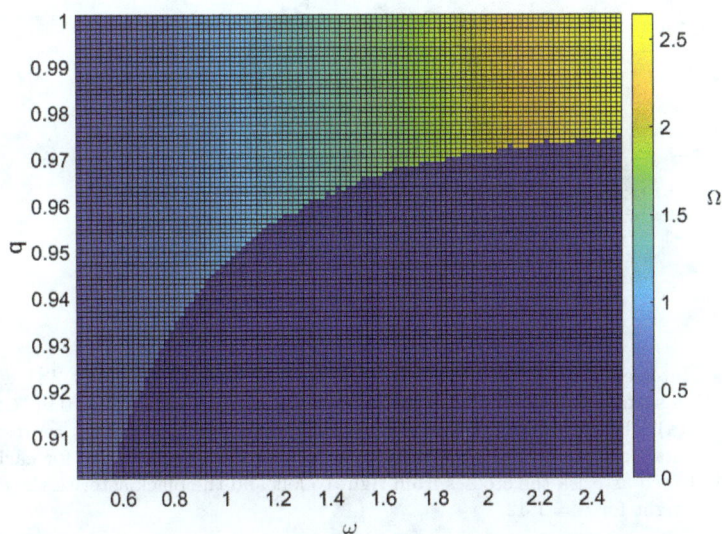

Fig. 4.12   Frequency response of Eq. (4.31) Ω to changes in q and w.

done with the Poincaré planes, although in both cases many crossing points were lost, is important to remark that for the fractional case, the return points stay aligned with the return points of the multi-scroll attractor i.e. the multi-scroll attractor and the fractional multi-stable attractor share crossing points.

(a)

(b)

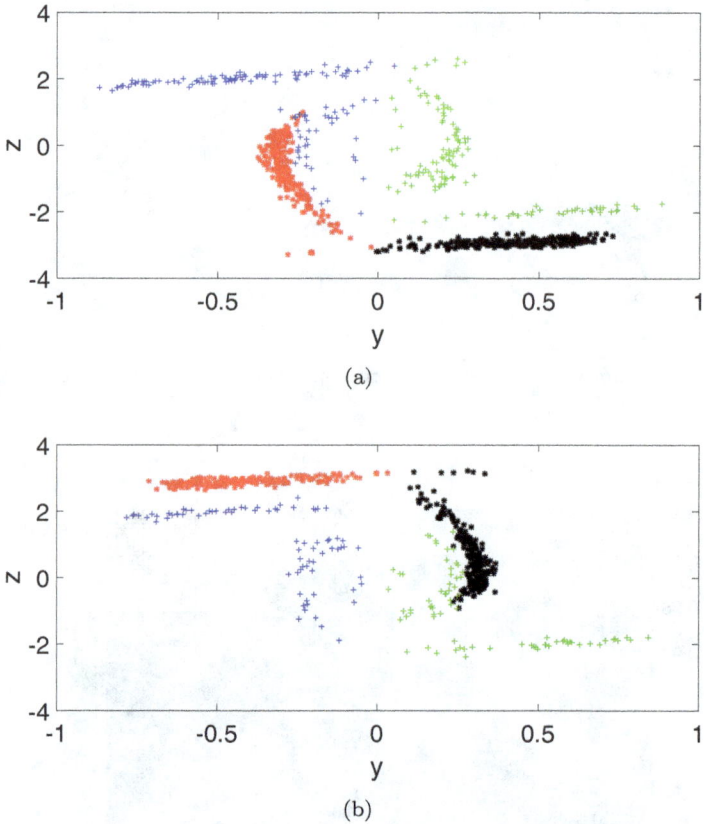

Fig. 4.13   Intersections of the trajectory of the system (4.30) with (4.28) with the commutation surface S_1 and S_2, for $a_1 = 10.5$; $a_2 = 7$; $a_3 = 0.7$; $C1 = 0.9$; $C2 = 0.6$ and $q = 1$ (a), (b) respectively. Marked in green cross the points that represent the crossing points from left to right and in blue cross the from right to left for each plane for $\alpha = 1$. In red asterisk the crosses from right to left and the black asterisk the crossing from left to right for $\alpha = 1.42$.

The crossing events of interest are to identify the crossing point, or in other words, the escape points and return points to the origin domain $\mathcal{D}_i$ to $\mathcal{D}_{i+1}$ or $\mathcal{D}_i$ to $\mathcal{D}_{i-1}$. In Figs. 4.14(a), (b) it is shown the trajectory intersections for the system (4.30) with $\alpha = 1.0$ and $q = 0.945$ for the right and left plane, respectively, in red asterisk the crosses from right to left and the black asterisk the crossing from left to right, in green and blue crosses the crossing points for $q = 1.0$ described in the Section 4.4.

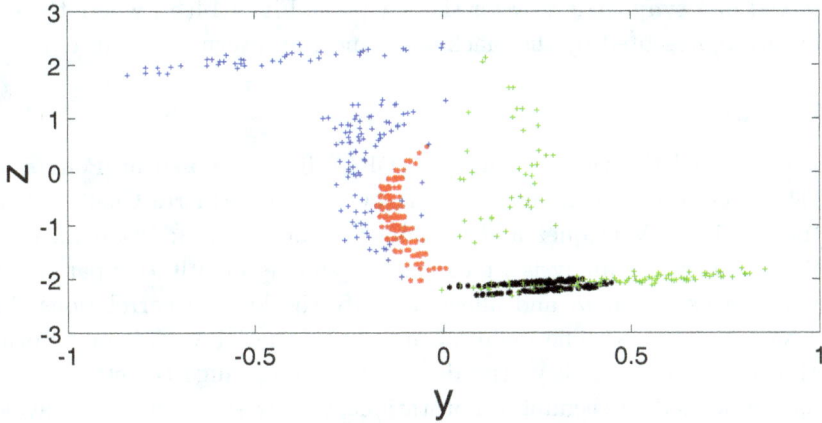

(a)

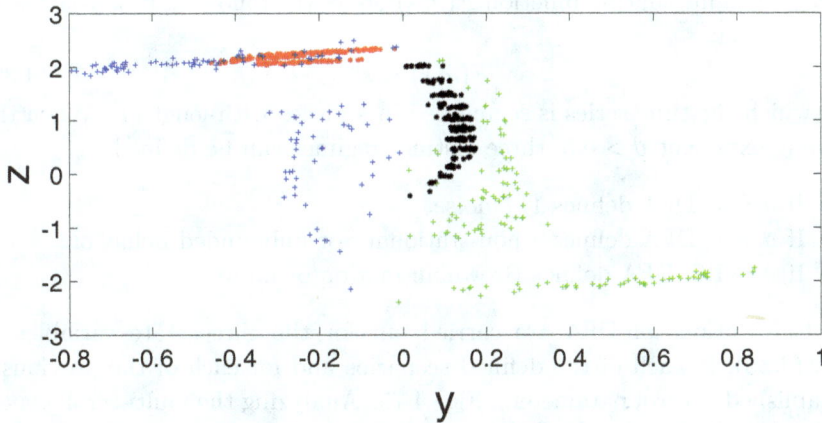

(b)

Fig. 4.14 Intersections of the trajectory of the system (4.30) with (4.28) with the commutation surface S_1 and S_2, for $a_1 = 10.5$; $a_2 = 7$; $a_3 = 0.7$; $C1 = 0.9$; $C2 = 0.6$ and $\alpha = 1$ (a), (b) respectively. Marked in green cross the points that represent the crossing points from left to right and in blue cross the from right to left for each plane for $q = 1$. In red asterisk the crosses from right to left and the black asterisk the crossing from left to right for $q = 0.945$.

Notice that for the right plane Fig. 4.14(a) the exit points of this domain are depicted in red, and compared with the multi-scroll attractor ($q = 1.0$), many crossing points were lost, as well as for the return points depicted in black many crossing points were lost. This same characteristic can be observed in a symmetric way for the left plane Fig. 4.14(b), where the exit point are represented by the black color and the return points in red.

4.5.3   *Correlation*

The Detrended Fluctuation Analysis (DFA), first proposed in [Peng *et al.* (1994)], and used to analyze chaotic systems in [Huerta-Cuellar *et al.* (2014b); Gilardi-Velázquez and Campos-Cantón (2018b); Pisarchik *et al.* (2018)], is a scaling analysis which yields a simple quantitative parameter, the scaling exponent ϑ, and allows identify the kind of correlations that the system presents. The main advantages of the DFA, over many other methods, are that it allows the detection of long-range correlations of a signal embedded in seemingly nonstationary time series, and also avoids the spurious detection of apparent long-range correlations that are an artifact of nonstationarity, infering dynamical properties from the temporal series. The fluctuation function $F(\nu; s)$ obeys the following power-law scaling relation:

$$F(\nu; s) \sim s^{\vartheta}, \tag{4.32}$$

for which the time series is segmented in s pieces with length ν. When the scaling exponent $\vartheta > 0.5$, three distinct regimes can be defined:

- If $\vartheta \sim 1$, DFA defines $1/f$ noise.
- If $\vartheta > 1$, DFA defines a non-stationary or unbounded behavior.
- If $\vartheta \sim 1.5$, DFA defines Brownian motion or noise.

Calculations of DFA are carried out for the three state variables of Eq. (4.30), in each of the defined scenarios and for each of the previously established control parameters, Fig. 4.15. Analyzing the multi-scroll generator with fractional-order-derivatives, causes a decrease in the DFA values, indicating that the dynamics of the system losses long-range correlations, approaching to values very close to pink noise ($\frac{1}{f}$) [Peng *et al.* (1994)]. The scenario with more notorious variations, in comparison with the integer-order behavior, is presented when the system is analyzed with a commensurate fractional-order ($S2$), where the dynamical response presents a lost in the correlations for $0.4 < \alpha < 0.65$, and the system recovers the original correlation values for $0.65 \leq \alpha \leq 0.75$.

The DFA study applied to y and z variables, shows a linear-constant loss of correlations, among the α increase. In contrast to the entire-order response, there are no significant changes. This implies that the most representative modifications are exhibited in the x variable. The results shown

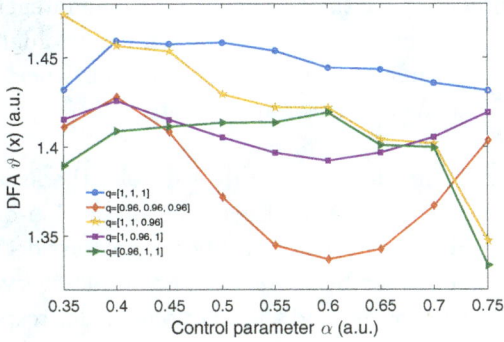

(a)

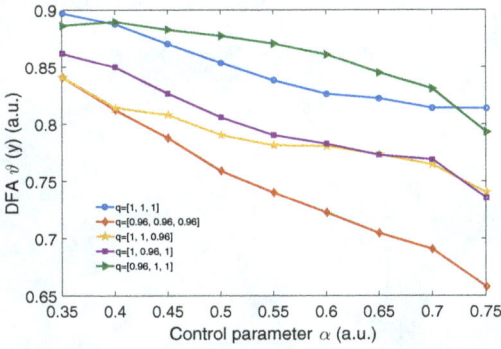

(b)

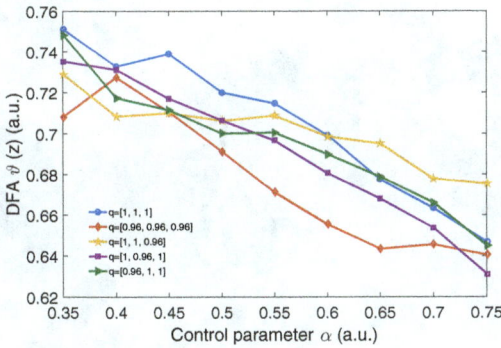

(c)

Fig. 4.15   DFA analysis for all the state variables considering the five defined scenarios. Applied to state variable (a) x, (b) y, and (c) z.

in Fig. 4.15(a), indicate that the response of $S1$, the integer-order system, has values in DFA very close to $\vartheta \sim 1.5$, which defines highly random behavior, characteristic value of the Brownian motion [Huerta-Cuellar *et al.* (2014b); Gilardi-Velázquez and Campos-Cantón (2018b); Pisarchik *et al.* (2018)].

Based on the α values for the x state variable, the system loses long-range correlations when considering fractional integration-orders being the case S3 the most significant since it presents the greatest decrease in scaling exponent. The integer-order behavior presents values very close to those expected for Brownian motion, and the fractional-order responses are more similar to pink-noise. The analysis based on DFA in the state variables y and z, does not present significant changes, implying that the system, no matter if analyzed in a commensurate or incommensurate scenario, is only affected in the x state.

Bibliography

Amaral, G., Letellier, C., and Aguirre, L. (2006). Piecewise affine models of chaotic attractors: The Rössler and Lorenz systems, *Chaos: An Interdisciplinary Journal of Nonlinear Science* **16**, doi:10.1063/1.2149527.

Andrievsky, B., Kuznetsov, N., Leonov, G., and Pogromsky, A. (2013). Hidden oscillations in aircraft flight control system with input saturation, *IFAC Proceedings Volumes* **46**, 12, pp. 75 – 79, doi:https://doi.org/10.3182/20130703-3-FR-4039.00026, `http://www.sciencedirect.com/science/article/pii/S1474667015339380`, 5th IFAC Workshop on Periodic Control Systems.

Angeli, D. (2007). Multistability in systems with counter-clockwise input-output dynamics, *IEEE Transactions on Automatic Control* **52**, 4, pp. 596–609, doi:10.1109/TAC.2007.894507.

Astakhov V, Shabunin A, Uhm W, K. S. (2001). Multistability formation and synchronization loss in coupled Hénon maps: two sides of the single bifurcational mechanism. *Physical Review E - Statistical, Nonlinear, and Soft Matter Physics* **63**, 5, p. 056212.

C. R. Hens, R. Banerjee, U. Feudel and Dana, S. K. (2012). How to obtain extreme multistability in coupled dynamical systems, *Phys. Rev. E* **85**, 3, p. 035202.

Cafagna, D. and Grassi, G. (2003). Hyperchaotic coupled chua circuits: An approach for generating new n × m-scroll attractors, *International Journal of Bifurcation and Chaos* **13**, 09, pp. 2537–2550, doi:10.1142/S0218127403008065, `https://doi.org/10.1142/S0218127403008065`, `https://doi.org/10.1142/S0218127403008065`.

Campos-Cantón, E. (2016). Chaotic attractors based on unstable dissipative systems via third-order differential equation, *International Journal of Modern Physics C* **27**, 01, p. 1650008, doi:10.1142/S012918311650008X.

Campos-Cantón, E., Barajas-Ramirez, J., Solis-Perales, G., and Femat, R. (2010). Multiscroll attractors by switching systems, *Chaos: An Interdisciplinary Journal of Nonlinear Science* **20**, 1, p. 013116.

Carvalho, R., Fernandez, B., and Vilela Mendes, R. (2001). From synchronization to multistability in two coupled quadratic maps, *Physics Letters, Section A: General, Atomic and Solid State Physics* **285**, 5-6, pp. 327–338, doi: 10.1016/S0375-9601(01)00370-X, `arXiv:0005053 [nlin.CD]`.

Chua, L. O., Shilnikov, L. P., Shilnikov, A. L., and Turaev, D. V. (2001). *Methods Of Qualitative Theory In Nonlinear Dynamics (Part II)*, Vol. 5 (World Scientific).

Connolly, J. A. and Ford, N. J. (2006). Comparison of numerical methods for fractional differential equations, *Communications on Pure & Applied Analysis* **5**, 2, p. 289.

Deng, W. and Lü, J. (2007). Generating multi-directional multi-scroll chaotic attractors via a fractional differential hysteresis system, *Physics Letters A* **369**, 5-6, pp. 438–443.

Diethelm, K. (2010). *The analysis of fractional differential equations: An application-oriented exposition using differential operators of Caputo type* (Springer Science & Business Media), doi:10.1007/978-3-642-14574-2.

Diethelm, K., Ford, N., and Freed, A. (2002). A predictor-corrector approach for the numerical solution of fractional differential equations, *Nonlinear Dynamics* **29**, 1-4, pp. 3–22, doi:10.1023/A:1016592219341.

Dudkowski, D., Jafari, S., Kapitaniak, T., Kuznetsov, N. V., Leonov, G. A., and Prasad, A. (2016). Hidden attractors in dynamical systems, *Physics Reports* **637**, pp. 1 – 50, doi:https://doi.org/10.1016/j.physrep.2016.05.002, `http://www.sciencedirect.com/science/article/pii/S0370157316300928`, hidden Attractors in Dynamical Systems.

Echenausía-Monroy, J., García-López, J., Jaimes-Reátegui, R., López-Mancilla, D., and Huerta-Cuellar, G. (2018). Family of bistable attractors contained in an unstable dissipative switching system associated to a snlf, *Complexity* **2018**, doi:10.1155/2018/6794791.

Echenausía-Monroy, J. and Huerta-Cuellar, G. (2019). A novel approach to generate attractors with a high number of scrolls, *Nonlinear Analysis: Hybrid Systems* **2019**, p. 100822, doi:10.1016/j.nahs.2019.100822.

Echenausía-Monroy, J. L., Huerta-Cuellar, G., Jaimes-Reátegui, R., García-López, J. H., Aboites, V., Cassal-Quiroga, B. B., and Gilardi-Velázquez, H. E. (2020). Multistability emergence through fractional-order-derivatives in a pwl multi-scroll system, *Electronics* **9**, 6, p. 880, doi:10.3390/electronics9060880.

Elwakil, A. S., Salama, K. N., and Kennedy, M. P. (2000). A system for chaos generation and its implementation in monolithic form, in *2000 IEEE International Symposium on Circuits and Systems (ISCAS)*, Vol. 5 (IEEE), pp. 217–220.

Escalante-González, R. J. and Campos, E. (2021). Multistable systems with nested hidden and self-excited double scroll attractors, *The European Physical Journal Special Topics* doi:10.1140/epjs/s11734-021-00350-3, `https://doi.org/10.1140/epjs/s11734-021-00350-3`.

Escalante-González, R. J. and Campos-Cantón, E. (2017). Generation of chaotic attractors without equilibria via piecewise linear systems, *International Journal of Modern Physics C* **28**, 01, p. 1750008, doi:10.1142/ S0129183117500085, `https://doi.org/10.1142/S0129183117500085`, `https://doi.org/10.1142/S0129183117500085`.

Escalante-González, R. J. and Campos-Cantón, E. (2019). A class of piecewise linear systems without equilibria with 3-D grid multiscroll chaotic attractors, *IEEE Transactions on Circuits and Systems II: Express Briefs* **66**, 8, pp. 1456–1460, doi:10.1109/TCSII.2018.2886526.

Escalante-González, R. J. and Campos-Cantón, E. (2019). Coexistence of hidden attractors and self-excited attractors through breaking heteroclinic-like orbits of switched systems, *arXiv e-Print archive*.

Escalante-González, R. J., Campos-Cantón, E., and Nicol, M. (2017). Generation of multi-scroll attractors without equilibria via piecewise linear systems, *Chaos: An Interdisciplinary Journal of Nonlinear Science* **27**, 5, p. 053109, doi:10.1063/1.4983523, `https://doi.org/10.1063/1.4983523`, `https://doi.org/10.1063/1.4983523`.

Escalante-González, R. J. and Campos, E. (2020a). Hyperchaotic attractors through coupling of systems without equilibria, *The European Physical Journal Special Topics* **229**, pp. 1309–1318, doi:10.1140/epjst/ e2020-900197-4, `https://doi.org/10.1140/epjst/e2020-900197-4`.

Escalante-González, R. J. and Campos, E. (2020b). Hyperchaotic attractors through coupling of systems without equilibria, *The European Physical Journal Special Topics* **229**, pp. 1309–1318, doi:10.1140/epjst/ e2020-900197-4, `https://doi.org/10.1140/epjst/e2020-900197-4`.

Escalante-González, R. J. and Campos, E. (2020c). Multistable systems with hidden and self-excited scroll attractors generated via piecewise linear systems, *Complexity*.

García-Martínez, M., Ontañón-García, L., Campos-Cantón, E., and Čelikovský, S. (2015). Hyperchaotic encryption based on multi-scroll piecewise linear systems, *Applied Mathematics and Computation* **270**, pp. 413–424, doi: https://doi.org/10.1016/j.amc.2015.08.037, `http://www.sciencedirect.com/science/article/pii/S0096300315010929`.

Gershman, S. J., Vul, E., and Tenenbaum, J. B. (2012). Multistability and Perceptual Inference, *Neural Computation* **24**, 1, pp. 1–24, doi:10.1162/ NECO_a_00226.

Ghaffarizadeh, A., Flann, N. S., and Podgorski, G. J. (2014). Multistable switches and their role in cellular differentiation networks. *BMC bioinformatics* **15 Suppl 7**, Suppl 7, p. S7, doi:10.1186/1471-2105-15-S7-S7, `http://www.pubmedcentral.nih.gov/articlerender.fcgi?artid=4110729{&}tool=pmcentrez{&}rendertype=abstract`.

Giesl, P. (2007). On the determination of the basin of attraction of discrete dynamical systems, *Journal of Difference Equations and Applications* **13**, 6, pp. 523–546, doi:10.1080/10236190601135209, `http://www.scopus.com/inward/record.url?eid=2-s2.0-34249683317{&}partnerID=tZOtx3y1`.

Gilardi-Velázquez, H. and Campos-Cantón, E. (2018a). Nonclassical point of view of the brownian motion generation via fractional deterministic model, *International Journal of Modern Physics C* **29**, 03, p. 1850020, doi: 10.1142/S0129183118500201.

Gilardi-Velázquez, H. and Campos-Cantón, E. (2018b). Nonclassical point of view of the brownian motion generation via fractional deterministic model, *International Journal of Modern Physics C* **29**, 03, p. 1850020, doi: 10.1142/S0129183118500201.

Gilardi-Velázquez, H., Ontañón-García, L., Hurtado-Rodriguez, D., and Campos-Cantón, E. (2016). Multistability in piecewise linear systems by means of the eigenspectra variation and the round function, *arXiv preprint arXiv:1611.03461*.

Gilardi-Velázquez, H., Ontañón-García, L., Hurtado-Rodriguez, D., and Campos-Cantón, E. (2017). Multistability in piecewise linear systems versus eigenspectra variation and round function, *International Journal of Bifurcation and Chaos* **27**, 09, p. 1730031, doi:10.1142/S0218127417300312.

Haddad, W. M., Hui, Q., and Bailey, J. M. (2011). Multistability, bifurcations, and biological neural networks: A synaptic drive firing model for cerebral cortex transition in the induction of general anesthesia, in *Proceedings of the IEEE Conference on Decision and Control*, ISBN 9781612848006, pp. 3901–3908, doi:10.1109/CDC.2011.6160350.

Hoover, W. G. (1995). Remark on "some simple chaotic flows", *Phys. Rev. E* **51**, pp. 759–760, doi:10.1103/PhysRevE.51.759, https://link.aps.org/doi/10.1103/PhysRevE.51.759.

Hu, X., Liu, C., Liu, L., Ni, J., and Li, S. (2016). Multi-scroll hidden attractors in improved Sprott A system, *Nonlinear Dynamics* **86**, 3, pp. 1725–1734, doi:10.1007/s11071-016-2989-5, https://doi.org/10.1007/s11071-016-2989-5.

Huerta-Cuellar, G., Jimenez-Lopez, E., Campos-Cantón, E., and Pisarchik, A. (2014a). An approach to generate deterministic brownian motion, *Communications in Nonlinear Science and Numerical Simulation* **19**, 8, pp. 2740–2746.

Huerta-Cuellar, G., Jimenez-Lopez, E., Campos-Cantón, E., and Pisarchik, A. (2014b). An approach to generate deterministic brownian motion, *Communications in Nonlinear Science and Numerical Simulation* **19**, 8, pp. 2740–2746, doi:10.1016/j.cnsns.2014.01.010.

Hui, Q. (2014). Multistability Analysis of Discontinuous Dynamical Systems via Finite Trajectory Length, in *World Automation Congress (WAC), 2014* (IEEE), doi:10.1109/WAC.2014.6935993.

Jafari, S. and Sprott, J. (2013). Simple chaotic flows with a line equilibrium, *Chaos, Solitons and Fractals* **57**, pp. 79–84, doi:https://doi.org/10.1016/j.chaos.2013.08.018, http://www.sciencedirect.com/science/article/pii/S096007791300177X.

Jafari, S., Sprott, J., and Golpayegani, S. M. R. H. (2013). Elementary quadratic chaotic flows with no equilibria, *Physics Letters A* **377**, pp. 699–702.

Jiménez-López, E., Salas, J. G., Ontañón-García, L. J., Campos-Cantón, E., and Pisarchik, A. N. (2013). Generalized multistable structure via chaotic synchronization and preservation of scrolls, *Journal of the Franklin Institute* **350**, 10, pp. 2853–2866.

Jung, P., Butz, S., Marthaler, M., Fistul, M. V., Leppäkangas, J., Koshelets, V. P., and Ustinov, a. V. (2014). Multistability and switching in a superconducting metamaterial. *Nature communications* **5**, p. 3730, doi: 10.1038/ncomms4730, `arXiv:1312.2937`, `http://www.ncbi.nlm.nih.gov/pubmed/24769498`.

Kaplan, J. L. and Yorke, J. A. (1979). Chaotic behavior of multidimensional difference equations, in *Functional differential equations and approximation of fixed points* (Springer), pp. 204–227.

Kengne, J. (2017). On the Dynamics of Chua's oscillator with a smooth cubic nonlinearity: occurrence of multiple attractors, *Nonlinear Dynamics* **87**, 1, pp. 363–375.

Kengne, J., Njitacke Tabekoueng, Z., and Fotsin, H. B. (2016). Coexistence of multiple attractors and crisis route to chaos in autonomous third order Duffing-Holmes type chaotic oscillators, *Communications in Nonlinear Science and Numerical Simulation* **36**, pp. 29–44, doi:10.1016/j.cnsns.2015.11.009.

Leonov, G., Kuznetsov, N., and Vagaitsev, V. (2011a). Localization of hidden chua's attractors, *Physics Letters A* **375**, 23, pp. 2230 – 2233, doi:https://doi.org/10.1016/j.physleta.2011.04.037, `http://www.sciencedirect.com/science/article/pii/S0375960111005135`.

Leonov, G., Kuznetsov, N., and Vagaitsev, V. (2011b). Localization of hidden Chua's attractors, *Physics Letters A* **375**, 23, pp. 2230–2233.

Li, C. and Sprott, J. C. (2013). Multistability in a Butterfly Flow, *International Journal of Bifurcation and Chaos* **23**, 12, p. 1350199, doi:10.1142/S021812741350199X, `http://www.worldscientific.com/doi/abs/10.1142/S021812741350199X`.

Li, C., Sprott, J. C., Thio, W., and Zhu, H. (2014). A new piecewise linear hyperchaotic circuit, *IEEE Transactions on Circuits and Systems—Part II: Express Briefs* **61**, 12, pp. 977–981.

Lu, J., Chen, G., Yu, X., and Leung, H. (2004). Design and analysis of multiscroll chaotic attractors from saturated function series, *IEEE Transactions on Circuits and Systems I: Regular Papers* **51**, 12, pp. 2476–2490.

Lü, J., Han, F., Yu, X., and Chen, G. (2004). Generating 3-d multi-scroll chaotic attractors: A hysteresis series switching method, *Automatica* **40**, 10, pp. 1677–1687, doi:10.1016/j.automatica.2004.06.001.

Mendes, R. V. (2000). Multistability in dynamical systems, in J. Gambaudo, P. Hubert, P. Tisseur, and S. Vaienti (eds.), *Dynamical Systems: From Crystal to Chaos. Proceedings of the conference in honor of Gerar Rauzy on his 60th birthday*, 1st edn. (World Scientific, Singapur), ISBN ISBN 981-02-4217-4, p. 105.113.

Monje, C. A., Chen, Y., Vinagre, B. M., Xue, D., and Feliu-Batlle, V. (2010). *Fractional-order systems and controls: fundamentals and applications* (Springer Science & Business Media).

Ontañón-García, L. and Campos-Cantón, E. (2017). Widening of the basins of attraction of a multistable switching dynamical system with the location of symmetric equilibria, *Nonlinear Analysis: Hybrid Systems* **26**, pp. 38–47, doi:https://doi.org/10.1016/j.nahs.2017.04.002, http://www.sciencedirect.com/science/article/pii/S1751570X17300262.

Ontañón-García, L. J. and Campos-Cantón, E. (2015). Displacement in space of the equilibria of unstable dissipative systems, in *7th International Scientific Conference on Physics and Control*, pp. 19–22.

Ontañón-García, L. J., Jiménez-López, E., Campos-Cantón, E., and Basin, M. (2014). A family of hyperchaotic multi-scroll attractors in rn, *Applied Mathematics and Computation* **233**, pp. 522–533.

Oseledec, V. I. (1968). A multiplicative ergodic theorem. Lyapunov characteristic number for dynamical systems, *Trans. Moscow Math. Soc.* **19**, pp. 197–231.

Patel, M. S., Patel, U., Sen, A., Sethia, G. C., Hens, C., Dana, S. K., Feudel, U., Showalter, K., Ngonghala, C. N., and Amritkar, R. E. (2014). Experimental observation of extreme multistability in an electronic system of two coupled Rössler oscillators, *Physical Review E - Statistical, Nonlinear, and Soft Matter Physics* **89**, 2, doi:10.1103/PhysRevE.89.022918.

Peng, C., Buldyrev, S., Havlin, S., Simons, M., Stanley, H., and Goldberger, A. (1994). Mosaic organization of dna nucleotides, *Physical review e* **49**, 2, p. 1685, doi:10.1103/physreve.49.1685.

Perko, L. (2013). *Differential equations and dynamical systems*, Vol. 7 (Springer Science & Business Media).

Petráš, I. (2011). *Fractional-order nonlinear systems: modeling, analysis and simulation* (Springer Science & Business Media).

Petrzela, J., Pospíšil, J., Kolka, Z., and Hanus, S. (2003). Lorenz and Rössler systems with piecewise-linear vector fields, *WSEAS Transactions on Mathematics* **2**.

Pisarchik, A., Huerta-Cuellar, G., and Kulp, C. (2018). Statistical analysis of symbolic dynamics in weakly coupled chaotic oscillators, *Communications in Nonlinear Science and Numerical Simulation* **62**, pp. 134–145, doi:10.1016/j.cnsns.2018.02.025.

Pisarchik, A. and Jaimes-Reátegui, R. (2015). Deterministic coherence resonance in coupled chaotic oscillators with frequency mismatch, *Physical Review E* **92**, 5, p. 050901.

Podlubny, I. (1998). *Fractional differential equations: an introduction to fractional derivatives, fractional differential equations, to methods of their solution and some of their applications*, Vol. 198 (Elsevier).

Rossler, O. (1979). An equation for hyperchaos, *Physics Letters A* **71**, 2, pp. 155–157, doi:https://doi.org/10.1016/0375-9601(79)90150-6, https://www.sciencedirect.com/science/article/pii/0375960179901506.

Rukhin, A. (2010). A statistical test suite for random and pseudo-random number generators for cryptographic applications, *NIST special publication*.

Sagués, F. and Epstein, I. R. (2003). Nonlinear chemical dynamics, *Dalton Transactions* , 7, pp. 1201–1217, doi:10.1039/b210932h.

Sevilla-Escoboza, R., Pisarchik, A. N., Jaimes-Reátegui, R., and Huerta-Cuellar, G. (2015). Selective monostability in multi-stable systems, *Proceedings of the Royal Society A: Mathematical, Physical and Engineering Sciences* **471**, 2180, p. 20150005.

Silva, C. P. (1993). Shil'nikov's theorem-a tutorial, *IEEE Transactions on Circuits and Systems I: Fundamental Theory and Applications* **40**, 10, pp. 675–682.

Sprott, J. C. (1994). Some simple chaotic flows, *The American Physical Society, Physical Review E* **50**, 2, pp. 647–650.

Tavazoei, M. S., Haeri, M., Bolouki, S., and Siami, M. (2009). Stability preservation analysis for frequency-based methods in numerical simulation of fractional order systems, *SIAM Journal on Numerical Analysis* **47**, 1, pp. 321–338.

Tresser, C. (1984). About some theorems by lp šil'nikov, in *Annales de l'IHP Physique théorique*, Vol. 40, pp. 441–461.

Wang, X. and Chen, G. (2013). Constructing a chaotic system with any number of equilibria, *Nonlinear Dyn* **71**, 3, pp. 429 – 436, doi:https://doi.org/10. 1007/s11071-012-0669-7, `https://doi.org/10.1007/s11071-012-0669-7`.

Wei, Z. (2011). Dynamical behaviors of a chaotic system with no equilibria, *Physics Letters A* **376**, pp. 102–108.

Wei, Z., Zhu, B., and Escalante-González, R. (2021). Existence of periodic orbits and chaos in a class of three-dimensional piecewise linear systems with two virtual stable node-foci, *Nonlinear Analysis: Hybrid Systems* **43**, p. 101114, doi:https://doi.org/10.1016/j.nahs.2021.101114, `https://www. sciencedirect.com/science/article/pii/S1751570X21001047`.

Wiggins, S. (2013). *Global bifurcations and chaos: analytical methods*, Vol. 73 (Springer Science & Business Media).

Wiggins, S., Wiggins, S., and Golubitsky, M. (2003). *Introduction to applied nonlinear dynamical systems and chaos*, Vol. 2 (Springer).

Xie, G., Chen, P., and Liu, M. (2008). Generation of multidirectional multiscroll attractors under the third-order jerk system, in *2008 International Symposium on Information Science and Engineering*, Vol. 1 (IEEE), pp. 145–149.

Yalçin, M. E., Suykens, J. A. K., and Vandewalle, J. (2002). Families of scroll grid attractors, *International Journal of Bifurcation and Chaos* **12**, 1, pp. 23–41.

Zambrano-Serrano, E., Campos-Cantón, E., and Muñoz-Pacheco, J. (2016). Strange attractors generated by a fractional order switching system and its topological horseshoe, *Nonlinear Dynamics* **83**, 3, pp. 1629–1641, doi: 10.1007/s11071-015-2436-z.

Zhang, X. and Wang, C. (2019). Multiscroll hyperchaotic system with hidden attractors and its circuit implementation, *International Journal of Bifurcation and Chaos* **29**, 09, p. 1950117, doi:10.1142/S0218127419501177.

Zhang, W., Zhou, S., Li, H., and Zhu, H. (2009). Chaos in a fractional-order rössler system, *Chaos, Solitons & Fractals* **42**, 3, pp. 1684–1691.

Z.T. Njitacke, J. kengne, H.B. Fotsin, A. Nguomkam Negou, D. T. (2016). Coexistence of multiple attractors and crisis route to chaos in a novel memristive diode bidge-based Jerk circuit, *Chaos, Solitons and Fractals: the interdisciplinary journal of Nonlinear Science, and Nonequilibrium and Complex Phenomena* **91**, 2016, pp. 180–197.

Index

9 789811 274114